Simon Broadley

Philippa Gardom Hulme

Sue Hocking

Mark Matthews

Jim Newall

Contents

B2 Part 1 Cells and the growing plant   12

B2 Part 2 Genes and proteins, inheritance, gene technology, and speciation   44

C2 Part 1 Structures, properties, and uses   88

How to use this book

Welcome to your AQA GCSE Additional Science course.
This book has been specially written by experienced teachers and examiners to match the 2011 specification.

On these two pages you can see the types of pages you will find in this book, and the features on them. Everything in the book is designed to provide you with the support you need to help you prepare for your examinations and achieve your best.

Unit openers

Specification matching grid: This shows you how the pages in the unit match to the exam specification for GCSE Additional Science, so you can track your progress through the unit as you learn.

Why study this unit: Here you can read about the reasons why the science you're about to learn is relevant to your everyday life.

You should remember: This list is a summary of the things you've already learnt that will come up again in this unit. Check through them in advance and see if there is anything that you need to recap on before you get started.

Opener image: Every unit starts with a picture and information on a new or interesting piece of science that relates to what you're about to learn.

Main pages

Learning objectives: You can use these objectives to understand what you need to learn to prepare for your exams. Higher Tier only objectives appear in pink text.

Key words: These are the terms you need to understand for your exams. You can look for these words in the text in bold or check the glossary to see what they mean.

Questions: Use the questions on each spread to test yourself on what you've just read.

Higher Tier content: Anything marked in pink is for students taking the Higher Tier paper only. As you go through you can look at this material and attempt it to help you understand what is expected for the Higher Tier.

Worked examples: These help you understand how to use an equation or to work through a calculation. You can check back whenever you use the calculation in your work.

Summary and exam-style questions

Every summary question at the end of a spread includes an indication of how hard it is. These indicators show which grade you are working towards. You can track your own progress by seeing which of the questions you can answer easily, and which you have difficulty with.

When you reach the end of a unit you can use the exam-style questions to test how well you know what you've just learnt. Each question has a grade band next to it.

→ E	Working towards Grade E
→ C	Working towards Grade C
→ A*	Working towards Grade A*
G–E	Grades G–E
D–C	Grades D–C
B–A*	Grades B–A*

Revision checklist: This is a summary of the main ideas in the unit. You can use it as a starting point for revision, to check that you know about the big ideas covered.

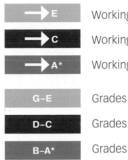

Visual summary: Another way to start revision is to use a visual summary, linking ideas together in groups so you can see how one topic relates to another. You can use this page as a start for your own summary.

Course catch-ups

Upgrade: Upgrade takes you through an exam question in a step-by-step way, showing you why different answers get different grades. Using the tips on the page you can make sure you achieve your best by understanding what each question needs.

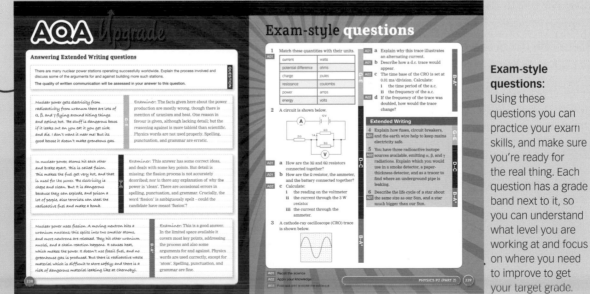

Exam-style questions: Using these questions you can practice your exam skills, and make sure you're ready for the real thing. Each question has a grade band next to it, so you can understand what level you are working at and focus on where you need to improve to get your target grade.

Routes and assessment

Matching your course

The units in this book have been written to match the specification, no matter what you plan to study after your GCSE Additional Science course.

In the diagram below you can see that the units and part units can be used to study either for **GCSE Additional Science**, or as part of **GCSE Biology**, **GCSE Chemistry** and **GCSE Physics** courses.

	GCSE Biology	GCSE Chemistry	GCSE Physics
GCSE Science	B1 (Part 1)	C1 (Part 1)	P1 (Part 1)
	B1 (Part 2)	C1 (Part 2)	P1 (Part 2)
GCSE Additional Science	**B2 (Part 1)**	**C2 (Part 1)**	**P2 (Part 1)**
	B2 (Part 2)	**C2 (Part 2)**	**P2 (Part 2)**
	B3 (Part 1)	C3 (Part 1)	P3 (Part 1)
	B3 (Part 2)	C3 (Part 2)	P3 (Part 2)

GCSE Additional Science assessment

The units in this book are broken into two parts to match the different types of exam paper on offer. The diagram below shows you what is included in each exam paper. It also shows you how much of your final mark you will be working towards in each paper.

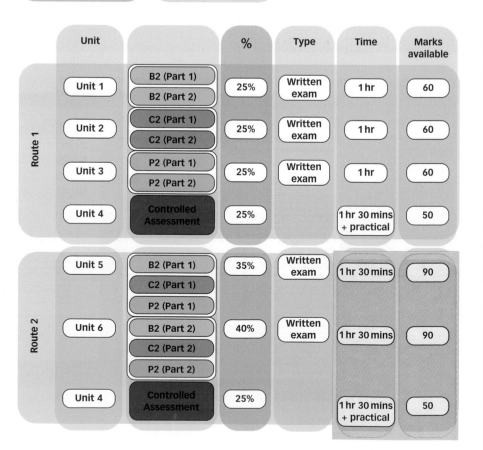

	Unit		%	Type	Time	Marks available
Route 1	Unit 1	B2 (Part 1) / B2 (Part 2)	25%	Written exam	1 hr	60
	Unit 2	C2 (Part 1) / C2 (Part 2)	25%	Written exam	1 hr	60
	Unit 3	P2 (Part 1) / P2 (Part 2)	25%	Written exam	1 hr	60
	Unit 4	Controlled Assessment	25%		1 hr 30 mins + practical	50
Route 2	Unit 5	B2 (Part 1) / C2 (Part 1) / P2 (Part 1)	35%	Written exam	1 hr 30 mins	90
	Unit 6	B2 (Part 2) / C2 (Part 2) / P2 (Part 2)	40%	Written exam	1 hr 30 mins	90
	Unit 4	Controlled Assessment	25%		1 hr 30 mins + practical	50

Understanding exam questions

When you read the questions in your exam papers you should make sure you know what kind of answer you are being asked for. The list below explains some of the common words you will see used in exam questions. Make sure you know what each word means. Always read the question thoroughly, even if you recognise the word used.

Calculate
Work out your answer by using a calculation. You can use your calculator to help you. You may need to use an equation; check whether one has been provided for you in the paper. The question will say if your working must be shown.

Describe
Write a detailed answer that covers what happens, when it happens, and where it happens. The question will let you know how much of the topic to cover. Talk about facts and characteristics. (Hint: don't confuse with 'Explain')

Explain
You will be asked how or why something happens. Write a detailed answer that covers how and why a thing happens. Talk about mechanisms and reasons. (Hint: don't confuse with 'Describe')

Evaluate
You will be given some facts, data or other information. Write about the data or facts and provide your own conclusion or opinion on them.

Outline
Give only the key facts of the topic. You may need to set out the steps of a procedure or process – make sure you write down the steps in the correct order.

Show
Write down the details, steps or calculations needed to prove an answer that you have been given.

Suggest
Think about what you've learnt in your science lessons and apply it to a new situation or a context. You may not know the answer. Use what you have learnt to suggest sensible answers to the question.

Write down
Give a short answer, without a supporting argument.

Top tips

Always read exam questions carefully, even if you recognise the word used. Look at the information in the question and the number of answer lines to see how much detail the examiner is looking for.

You can use bullet points or a diagram if it helps your answer.

If a number needs units you should include them, unless the units are already given on the answer line.

Controlled Assessment in GCSE Additional Science

As part of the assessment for your GCSE Additional Science course, you will undertake a Controlled Assessment task.

What is Controlled Assessment?

Controlled Assessment has taken the place of coursework for the new 2011 GCSE Science specifications. The main difference between coursework and Controlled Assessment is that you will be supervised by your teacher when you carry out your Controlled Assessment task.

What will my Controlled Assessment task look like?

Your Controlled Assessment task will be made up of four sections. These four sections make up an investigation, with each section looking at a different part of the scientific process.

	What will I need to do?	How many marks are available?
Research	• Independently develop your own hypothesis. • Research two methods for carrying out an experiment to test your hypothesis. • Prepare a table to record your results. • Carry out a risk assessment.	
Section 1	• Answer questions relating to your own research.	20 marks
Practical investigation	• Carry out your own experiment and record and analyse your results.	
Section 2	• Answer questions relating to the experiment you carried out. • Select appropriate data from data supplied by AQA and use it to analyse and compare with your hypothesis. • Suggest how ideas from your investigation could be used in a new context.	30 marks
	Total	**50 marks**

How do I prepare for my Controlled Assessment?

Throughout your course you will learn how to carry out investigations in a scientific way, and how to analyse and compare data properly.

On the next three pages there are Controlled Assessment-style questions matched to the biology, chemistry, and physics content in B2, C2, and P2. You can use them to test yourself, and to find out which areas you want to practise more before you take the Controlled Assessment task itself.

B2 Controlled Assessment-style questions

Overview: An increase in temperature may affect the rate of respiration. **You must develop your own hypothesis to test.** You will be provided with a culture of yeast cells, sugar solution, carbon dioxide and pH sensors, data loggers, temperature-controlled water baths, timers, rulers, and common laboratory glassware.

Download the Research Notes and Data Sheet for B2 from **www.oxfordsecondary.co.uk/ aqacasestudies**.

Research

*Record your findings in the **Research Notes table**.*

1. Research two methods to find out how temperature affects the rate of respiration.
2. Find out how the results might be useful in selecting the yeast used by breweries.

Section 1   Total 20 marks

Use your research findings to answer these questions.

1. (a) Name two sources that you used for your research.
   (b) Which of these sources did you find more useful, and why? [3]
2. (a) Write a hypothesis about how temperature may affect the rate of respiration in yeast cells.
   (b) Using information from your research, explain why you have developed this hypothesis. [3]
3. Describe how to carry out an investigation to test your hypothesis. Include the equipment needed and how to use it, the measurements to make, how to make it a fair test, and a risk assessment. [9]
4. Use your research to outline another possible method, and explain why it was not chosen. [3]
5. Draw a table to record data from the investigation. You may use ICT if you wish. [2]

Section 2   Total 30 marks

*Use the **Data Sheet** to answer these questions.*

1. Display the **Group A data** on a graph. *This data has been provided for you to use instead of data that you would gather yourself.* [4]
2. (a) What conclusion can you draw from the **Group A data** about a link between temperature and rate of respiration? Use any pattern you can see in the **Group A data** and quote figures from it. [3]
   (b) (i) Compare the **Group A** and **Group B data**. Do you think the **Group A data** is reproducible? Explain why. *The **Group B data** has been provided for you to use instead of data that would be gathered by others in your class.* [3]
      (ii) Explain how you could use the repeated results from Group B to obtain a more accurate answer. [3]
   (c) Look at the **Group A data**. Are there any anomalous results? Quote from the data. [3]
3. (a) Sketch a graph of **Case study 1** results. [2]
   (b) Explain to what extent the data from **Case studies 1–3** support or contradict your hypothesis. [3]
   (c) Compare the **Group A data** to the data shown on the **Case study 4** graph. Explain how far the **Case study 4** data supports or contradicts your hypothesis. [3]
4. In brewing, yeast respires anaerobically and produces ethanol as well as carbon dioxide. A company has developed and tested a new strain of yeast, strain X, for brewing strong (high alcohol content) beer. Their hypothesis is that strain X is better than strains Y and Z.
   (a) Does the data from **Case study 4** support their hypothesis? Quote from the data. [3]
   (b) The tests show that strain X has a different optimum temperature for growth than the other two strains. Heating yeast uses fuel and costs money. Show how ideas from the **Group A data** and the **Case studies** could be used by brewers. [3]

9

C2 Controlled Assessment-style questions

Overview: A change in temperature may cause a change in the rate of a reaction. **You must develop your own hypothesis to test.** You will be provided with magnesium ribbon, dilute hydrochloric acid, a thermometer, heating apparatus, and common laboratory glassware.

Download the Research Notes and Data Sheet for C2 from **www.oxfordsecondary.co.uk/aqacasestudies**.

Research

*Record your findings in the **Research Notes table**.*

1. Research two methods to find out whether temperature affects the reaction rate of acids.
2. Find out how the investigation results might be useful in choosing the best temperature at which to use a new limescale remover.

Section 1   Total 20 marks

Use your research findings to answer these questions.

1. **(a)** Name the two most useful sources that you used for your research.
   (b) Explain why these sources were the most useful. [3]
2. Write a hypothesis about how temperature might affect reaction rate. Use your research findings to explain why you made this hypothesis. [3]
3. Describe how to carry out an investigation to test your hypothesis. Include the equipment needed and how to use it, the measurements to make, how to make it a fair test, and a risk assessment. [9]
4. Use your research to outline another possible method, and explain why you did not choose it. [3]
5. Draw a table to record data from the investigation. You may use ICT if you wish. [2]

Section 2   Total 30 marks

*Use the **Data Sheet** to answer these questions.*

1. Display the **Group A data** on a graph. *This data has been provided for you to use instead of data that you would gather yourself.* [4]
2. **(a)** What conclusion can you draw from the **Group A data** about a link between temperature and reaction rate? Use any pattern you can see in the **Group A data** and quote figures from it to support your answer. [3]
   (b) (i) Compare the **Group A** and **Group B data**. Do you think the **Group A data** is reproducible? Explain why. *The Group B data has been provided for you to use instead of data that would be gathered by others in your class.* [3]
   (ii) Explain how you could use the repeated results from Group B to obtain a more accurate answer. [3]
   (c) Look at the **Group A data**. Are there any anomalous results? Quote from the data to explain your answer. [3]
3. **(a)** Sketch a graph of the results in **Case study 1**. [2]
   (b) Explain to what extent the data from **Case studies 1–3** support or contradict your hypothesis. [3]
   (c) Compare the **Group A data** to the data shown on the **Case study 4** graph. Explain how far the **Case study 4** data supports or contradicts your hypothesis. [3]
4. A company is developing a new limescale remover. It needs to choose the best temperature for its use. A chemist develops a hypothesis that the higher the temperature, the faster the reaction, and the quicker limescale is removed.
   (a) Does the **Group A data** support or contradict this hypothesis? Quote figures from the data to explain your answer. [3]
   (b) Show how ideas from the **Group A data** and the **Case studies** could be used by the company. [3]

P2 Controlled Assessment style questions

Overview: A force acting on an object may cause a change in shape of the object. When a force is applied to a spring it changes its length. **You must develop your own hypothesis to test.** You will be provided with a retort stand, clamp, boss, a small spring, 100 g masses, eye protection, and a metre rule.

Download the Research Notes and Data Sheet for P2 from **www.oxfordsecondary.co.uk/aqacasestudies**.

Research

*Record your findings in the **Research Notes table**.*

1. Research two different methods to find out how the force applied to a spring changes its length.
2. Find out how the results of the investigation might be useful in designing a simple force meter to measure the weight of rock samples.

Section 1   Total 20 marks

Use your research findings to answer these questions.

1. (a) Name the two most useful sources that you used for your research.
   (b) Explain why these sources were the most useful. [3]
2. Write a hypothesis about how the force applied to a spring might affect its extension. Use your research findings to explain why you made this hypothesis. [3]
3. Describe how to carry out an investigation to test your hypothesis. Include the equipment needed and how to use it, the measurements to make, how to make it a fair test, and a risk assessment. [9]
4. Use your research to outline another possible method, and explain why you did not choose it. [3]
5. Draw a table to record data from the investigation. You may use ICT if you wish. [2]

Section 2   Total 30 marks

*Use the **Data Sheet** to answer these questions.*

1. Display the **Group A data** on a graph. *This data has been provided for you to use instead of data that you would gather yourself.* [4]
2. (a) What conclusion can you draw from the **Group A data** about a link between force and extension? Use any pattern you can see in the **Group A data** and quote figures from it. [3]
   (b) (i) Compare the **Group A** and **Group B data**. Do you think the **Group A data** is reproducible? Explain why. *The **Group B data** has been provided for you to use instead of data that would be gathered by others in your class.* [3]
      (ii) Explain how you could use the repeated results from Group B to obtain a more accurate answer. [3]
   (c) Look at the **Group A data**. Are there any anomalous results? Quote from the data. [3]
3. (a) Sketch a graph of the results in **Case study 1**. [2]
   (b) Explain to what extent the data from **Case studies 1–3** support or contradict your hypothesis. [3]
   (c) Compare the **Group A data** to the **Case study 4 data**. Explain how far the **Case study 4 data** supports or contradicts your hypothesis. [3]
4. A spring manufacturing company claims that force applied to a spring is directly proportional to its extension up to a certain point.
   (a) Does the **Group A data** support or contradict this hypothesis? Quote figures from the data to explain your answer. [3]
   (b) Suggest how the results of your investigation and the **Case studies** might be useful in designing a simple force meter to measure the weight of rock samples. [3]

B2 Part 1

Cells and the growing plant

Why study this unit?

Photosynthesis is one of the most important biological processes. It is through photosynthesis that energy is trapped into the living world. Once trapped, this energy is used to power the entire living world in all its glory.

In this unit you will study photosynthesis as a process, and look at where the process occurs. You will also look at processes for sampling the distribution of living organisms in the environment.

The cell is often considered to be the basic building block of living things. This unit looks at the structure of plant and animal cells. You will observe how cells develop special structures to perform special functions. The principles of how cells work together to form organs, and how organs work together in organ systems, will form part of your study of cells. Finally, the problem of how molecules can get into and out of cells will be explored.

You should remember

1   You are made of cells that are organised into tissues, organs, and systems – such as the reproductive system.

2   Plant and animal cells both have a membrane, cytoplasm, nucleus, mitochondria, and ribosomes, but plant cells also have a cell wall and a large vacuole.

3   Cell structure and function.

4   The environment can be studied by sampling the distribution of organisms.

5   Plants make food by the process of photosynthesis.

6   Photosynthesis occurs in the leaves.

7   Diffusion is the movement of particles from a high concentration to a low concentration.

Have you ever considered how garden centres and nurseries get plants ready for sale at just the right time? Just how do poinsettia plants reach their stunning best in time for Christmas? It's all down to controlling plant growth.

The poinsettia has become a symbol of Christmas. In 2009, over 4 million British-grown plants were sold, plus many more imported plants. The production process for these plants starts in late August. Large greenhouses are used to grow them in. Shoots are pinched out to keep the plants short and full. To achieve the characteristic red colour, the plants are exposed to shortened days of less than 12 hours' light, with no light during the night. To produce strong, healthy plants, the daylight must be very bright, and the temperature must be kept above 10 °C, to promote the maximum rate of photosynthesis. The result will be strong, bushy plants, with glorious red colour.

1: Plant and animal cells

Learning objectives

After studying this topic, you should be able to:

✔ know that plants and animals are built out of cells

✔ know the parts of cells and their functions

Key words

cell, microscope, nucleus, cytoplasm, cell membrane, cell wall, chloroplast, permanent vacuole

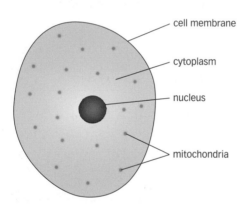

- cell membrane
- cytoplasm
- nucleus
- mitochondria

▲ A typical animal cell

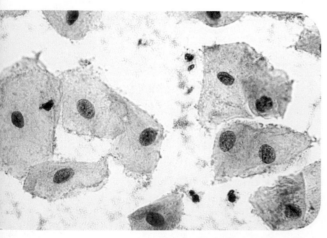

▲ Human cheek cells as seen through a microscope (× 2600)

Building blocks

All living things are made of one or more **cells**. Cells are the building blocks of living things. The larger the organism, the more cells it will contain. Usually, cells are very small and can only be seen using a **microscope**. Cells were first seen by Robert Hooke in 1665.

▲ The type of microscope used by Robert Hooke

▲ Coloured enhanced image of the cells of bark seen by Robert Hooke

Cell structure
Typical animal cells

Animal cells come in many different types, but they have certain features in common.

Cell part	Description and function
Nucleus	A large structure inside the cell. It contains chromosomes made of DNA. The nucleus controls the activities of the cell, and how it develops.
Cytoplasm	A jelly-like substance containing many chemicals. Most of the chemical reactions of the cell occur here.
Cell membrane	A thin layer around the cell. It controls the movement of substances into and out of the cell.
Mitochondria	Small rod-shaped structures that release energy from sugar during aerobic respiration.
Ribosomes	Small ball-shaped structures in the cytoplasm, where proteins are made.

Q:3 Not all plant cells contain chloroplasts because the root cell's cannot get the sunlight so don't need it.

A Explain why cells could not be seen until the development of the microscope. (too small)

B Which part of a cell controls the activities of the cell? (Nucleus)

C As you grow, what happens to the number of cells in your body? (multiply) multi cellur

Did you know...?

The cellulose in a plant cell wall helps support the plant, but it is also very useful to us. It makes the paper you are holding, and it is the fibre in our diet.

Typical plant cells

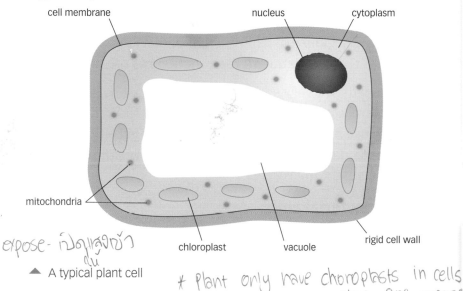

cell membrane · nucleus · cytoplasm

mitochondria · chloroplast · vacuole · rigid cell wall

▲ A typical plant cell

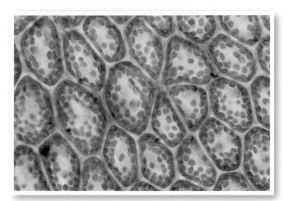

▲ Plant leaf cells seen through a microscope (× 500)

expose- เปิดเผยออกว่า
ดู

Plant cells have all of the structures seen in animal cells. But they also have one or two extra parts.

* Plant only have chloroplasts in cells that are exposed to sunlight

Exam tip · AQA

✔ Remember the difference between the cell membrane and the cell wall. Students often confuse these.

Cell part	Function
Cell wall	A layer outside the cell membrane. It is made of cellulose, which is strong and supports the cell.
Chloroplasts	Small discs found in the cytoplasm. They contain the green pigment chlorophyll. Chlorophyll traps light energy for photosynthesis.
Permanent vacuole	A fluid-filled cavity. The liquid inside is called cell sap. The sap helps support the cell.

If doesn't have in animal cell
มีสี
Sap-P สีเหลืองเหนียวในพืช

A group of organisms called algae are closely related to plants. They include seaweeds. Algal cells have a cell structure exactly like that of plant cells.

Questions

1 What is the function of the cell membrane? control the movement out side into & the cell

2 Plants are not as flexible as animals. Can you suggest a reason why? Because the wall of thin cell made from Cellulose

3 Not all plant cells contain chloroplasts. Suggest why root cells do not contain chloroplasts. Ground plane don't have light.

(เวก-กัด-โกม)
Vacuole is the small carity in the cytoplasm of a cell, bound by a single membrane and containing water, food, or metobolic waste.

2: Bacterial and fungal cells

Learning objectives

After studying this topic, you should be able to:

- ✔ know the structure of bacterial cells
- ✔ know the structure of fungal cells

Key words

bacteria, fungi

Other types of cell

As well as animals and plants, two other important groups of organism are bacteria and fungi. These organisms are built out of cells that have a slightly different structure from animal or plant cells.

Bacterial cells

Bacteria are a large group of organisms. Some are useful to us, for example, for breaking down waste or making food, whilst others cause problems such as diseases. Bacteria are single-celled organisms. At first sight bacterial cells look simple, but they carry out all the functions of other cells.

Bacterial cells are very small and can only just be seen using a light microscope. To see the detail of bacterial cells, biologists use high-powered microscopes called electron microscopes. These microscopes magnify thousands of times more than a light microscope.

Typical bacterial cells

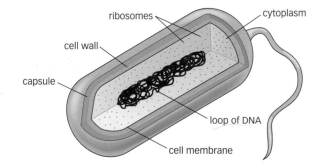

▲ A typical bacterial cell

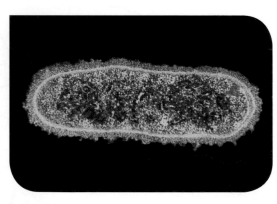

▲ The bacterium *E. coli* seen under an electron microscope (× 15 000)

Bacterial cells all share some features in common. Some parts are similar to those of plant and animal cells.

Cell part	Function
Cytoplasm	A jelly-like substance where most of the cell's reactions occur.
Cell membrane	Controls the movement of molecules into and out of the cell.
Cell wall	Having the same function as in a plant cell of maintaining the shape of the cell, but made of a different chemical instead of cellulose.
Ribosomes	Make proteins.

Bacterial cells also have parts not found in plant and animal cells.

Cell part	Function
Loop of DNA ıɪɪɪ٥ (nucleus)	DNA which controls the cell, as bacterial cells do not have a nucleus.
Capsule	Some bacteria have a slimy capsule around the outside of the cell wall, which protects them against antibiotics, for example.

Fungal cells

Fungi are another important group of organisms. They include mushrooms, moulds, and yeasts. Yeasts are commercially useful to us in the making of bread and beers. Yeasts are single-celled, but larger than a bacterial cell. They can be clearly seen using a light microscope, but the detail can be seen more clearly using an electron microscope.

Typical fungal cells ๆ เซลล์หมด (Typ)

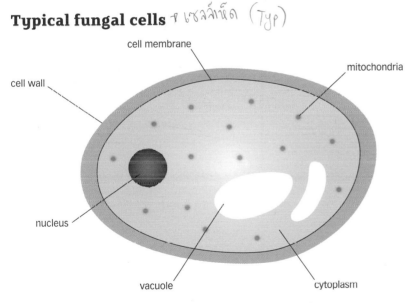

▲ A typical fungal cell

Fungal cells have many parts in common with other cells. The fungal cell has a membrane, cytoplasm and a nucleus, which function as they do in plant cells. The fungal cell wall is similar to that of a plant or bacterial cell. It has the same function, but is made of a third chemical called chitin.

A Explain why we need to use powerful electron microscopes to see bacteria.

B Which part of a bacterial cell performs the function of a nucleus?

C Name two parts of a bacterial cell that have the same structure and function as they do in an animal cell.

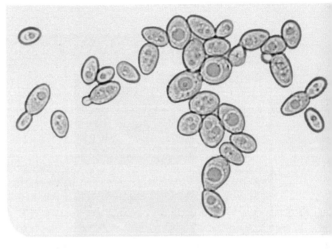

▲ Baker's yeast seen under a powerful light microscope (×750)

Questions

1 State one way in which a fungal cell and a bacterial cell are (a) similar and (b) different.

2 How could you tell a bacterial cell and a plant cell apart?

3 Explain why biologists discovered bacterial cells much later than plant cells.

E

C

A*

3: Specialised cells

Learning objectives

After studying this topic, you should be able to:

✔ know that cells can become specialised to perform different functions

Differentiation

Cells all have the same basic structure. However, not all cells end up having the same function. They become **specialised** to carry out their particular job. This is called **differentiation**. In becoming specialised, the cell may develop particular structures or become a specific shape. These new structures or shapes help the cell perform its function efficiently.

Specialised animal cells

Here are some examples of how animal cells may become specialised. These types of cell are all found in humans.

Cell type	Specialised structure	Function of cell
Red blood cell	Lacks a nucleus. Large surface area. Cell is small so fits into narrowest blood vessels. Contains haemoglobin which binds reversibly to oxygen.	Haemoglobin binds to oxygen and transports it around the body. The red blood cell gives up the oxygen to other body cells that need it.
Nerve cell	Many short extensions at the ends of the nerve. One long nerve fibre extension. Nerve fibre insulated with fatty sheath.	Receives impulses from other nerve cells via its many extensions. The impulses travel along the long nerve fibre. The insulation prevents loss of the impulse and makes it travel quickly.
Muscle cell	Cell is long and thin. Full of proteins that can make it contract.	The contractile proteins shorten the cell. This brings about movement.
Sperm cell	Cell has a head containing a nucleus, and a long tail.	The tail helps the cell to swim to the egg. The nucleus contains DNA which combines with the DNA of the egg cell.
Ciliated epithelial cell	Tall column-shaped cells. Cells can pack tightly together. Each cell covered at the top with fine hairs called cilia.	Tightly packed cells form a covering layer of cells. The cilia beat to create a current which can move particles such as bacteria up and out of the windpipe.

A Why are there many different types of cell in the human body?

B Explain how the following cells use their specialised structure to carry out their function:
   (a) red blood cells   (b) nerve cells

C Explain why a red blood cell cannot carry out the function of a muscle cell.

Key words

specialised, differentiation

Specialised plant cells

Cell type		Specialised structure	Function of cell
Palisade mesophyll cell		Found in the upper part of the leaf. Column-shaped cells with many chloroplasts.	The shape means that many cells can pack side by side. The chloroplasts contain chlorophyll for trapping light.
Root hair cell		Found in the young root. Long extension that protrudes out into the soil.	The extension increases the surface area of the cell, which improves its ability to absorb water and minerals from the soil.
Xylem		Found in roots, stems, and leaves. Hardened cell wall. Hollow inside with no living contents.	The hard cell wall gives strength, which helps support the plant. Being hollow allows the xylem to transport water.
Phloem		Found in roots, stems, and leaves. End walls of cells perforated. Cells largely hollow inside with small living cells next to them.	The hollow cavity and perforated end walls allow sugars to move through the plant. The living neighbouring cells supply energy for the transport of sugars.

Questions

1 Explain how the palisade mesophyll cell is adapted to its function of photosynthesis.

2 Why are xylem cells needed in the root, stem, and leaves of a plant? ↓ E

3 Explain how the xylem cell is specialised to carry out its functions. ↓ C

4 Explain how a large surface area helps the root hair cell to absorb water and minerals. ↓ A*

Learning objectives

After studying this topic, you should be able to:

✔ understand the process of diffusion

✔ know how diffusion allows particles to enter and leave cells

▲ Potassium permanganate crystals have been placed in a beaker of water, and after two hours the particles have diffused throughout the water

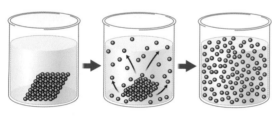

▲ Diffusion is the movement of particles along a concentration gradient

A Define diffusion.

B List some important molecules that diffuse into and out of cells by diffusion.

C Explain how the cell membrane can control which substances enter or leave the cell.

Getting in and out

Cells carry out many reactions. They need a constant supply of some substances, and need to get rid of others. So dissolved particles (molecules and ions) need to get into and out of cells. One important way that particles can move into or out of a cell is by **diffusion**.

Diffusion

Particles in a gas or in solution constantly move around. Particles tend to move from an area where they are in high concentration to an area where they are in lower concentration, along a **concentration gradient**. The particles move until they are evenly spread. This random movement of particles along a concentration gradient is called diffusion.

Diffusion in cells

Many dissolved substances enter and leave cells by diffusion, including important molecules like oxygen, which is needed for respiration in plant and animal cells. Carbon dioxide also gets into and out of cells by diffusion. Substances can diffuse as gases, or as dissolved particles in solution.

To get into a cell, particles pass through the cell membrane. The membrane will only allow small molecules through. This is fine for oxygen and carbon dioxide, as they are both small molecules. The process of diffusion does not use energy, because the molecules move spontaneously from regions of high concentration to regions of low concentration.

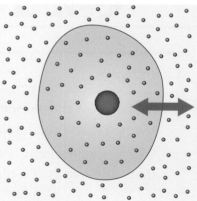

▲ Particles moving into cells by diffusion

Factors that affect the rate of diffusion

Diffusion happens because particles in solution constantly move. They can move in any direction, but far more particles tend to move from high to low concentration than the other way. This gives a net movement of particles from high concentration to low, along the concentration gradient. However, the rate of diffusion can vary. For example, increasing the temperature will give molecules more energy and the rate of diffusion is faster. There are several factors that can affect the rate of diffusion in cells.

Distance

The shorter the distance the particles have to move, the quicker the rate of diffusion will be. For example, if carbon dioxide has to reach cells in the centre of the leaf, then the thinner the leaf, the shorter the distance the gas has to travel and the quicker it will reach the cells.

Concentration gradient

The greater the difference in concentration between two regions, the faster the rate of diffusion. For example, leaf cells produce oxygen as a waste gas during photosynthesis. There is a build-up of oxygen in the leaf, giving a steep concentration gradient of oxygen between the inside and outside of the leaf. This leads to rapid diffusion of oxygen out of the leaf.

Surface area

The greater the surface area that the particles have to diffuse across, the quicker the rate of diffusion. For example, the lungs of animals and the internal structures of a leaf have a large surface area. This allows gases to diffuse rapidly into and out of cells.

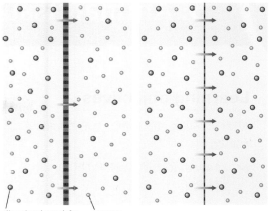

▲ The rate of diffusion depends on the distance the dissolved particles have to travel

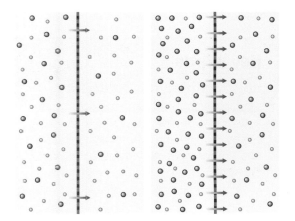

▲ The rate of diffusion depends on the concentration gradient

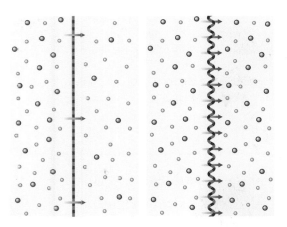

▲ The rate of diffusion depends on the surface area

Questions

1 State three factors that affect the rate of diffusion. **E**

2 Oxygen diffuses across the gills of a fish. Do you expect the cells lining the gills to be thick or thin? Explain why. **C**

3 Oxygen is constantly being used up in cells during respiration. Explain how this helps maintain the diffusion of oxygen into cells. **A***

Key words

diffusion, concentration gradient

Learning objectives

After studying this topic, you should be able to:

- ✔ know that cells work together as tissues
- ✔ know that tissues work together as organs
- ✔ appreciate the working of some animal tissues and organs

Key words

multicellular, tissue, organ

A What is a tissue?

B Why do multicellular animals have cells organised into tissues?

C What substances can glandular tissue produce?

Working together

Animals such as humans are built out of many cells – they are **multicellular**. Their cells do not work alone. It is more organised and efficient for the cells to work together.

You have seen that animal cells become specialised for a particular function. In the body of a multicellular animal, similar cells are organised together as a **tissue**. A tissue is a group of cells with a similar structure and function, working together. Organising cells into tissues allows life functions to be carried out more efficiently.

Animal tissues

Animal tissues include:

Muscle tissue

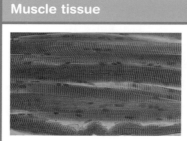

▲ Skeletal muscle

A group of muscle cells. When the cells contract together the entire muscle shortens, bringing about movement.

Glandular tissue

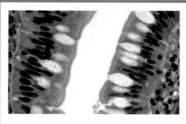

▲ Gut lining with goblet cells releasing liquid

A group of cells that produce a substance and then release it. Glandular tissue produces hormones in glands such as the pancreas, or enzymes in the cells lining the gut.

Epithelial tissue

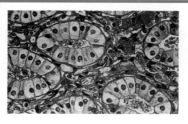

▲ Epithelial tissue lining the tubes of the kidney

A group of cells that form a covering layer for some parts of the body. These cover and protect parts of the body, or act as lining.

Organs

Tissues are grouped together to form **organs**. Organs are usually made of several different types of tissues. The tissues of an organ work together to perform one major task.

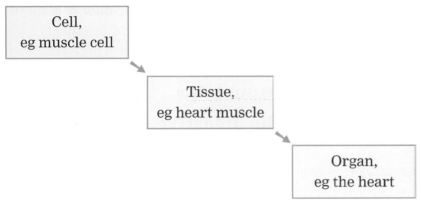

▲ Cells are organised into tissues and organs

The stomach

The stomach is an example of a human organ. It is made of several different tissues including muscle, glandular, and epithelial tissue.

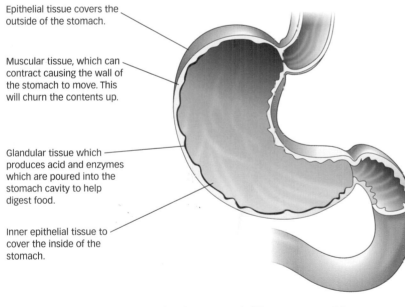

Epithelial tissue covers the outside of the stomach.

Muscular tissue, which can contract causing the wall of the stomach to move. This will churn the contents up.

Glandular tissue which produces acid and enzymes which are poured into the stomach cavity to help digest food.

Inner epithelial tissue to cover the inside of the stomach.

▲ The stomach is an organ that has several different types of tissue working together

Exam tip

✔ Understand the idea of scale here. Cells are small. Tissues are made of many cells, so they are larger, while organs are larger still.

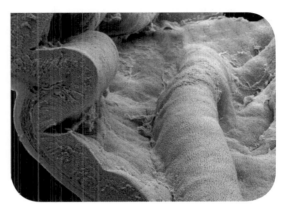

▲ A scanning electron micrograph of the stomach surface and a cross-section of the stomach wall (× 30)

Questions

1 What is the difference between a tissue and an organ?

2 Name three human organs other than the stomach.

3 Explain why tissues are larger than cells.

4 Explain how muscular tissue and glandular tissue work together to help bring about digestion of food in the stomach.

E

C

A*

Learning objectives

After studying this topic, you should be able to:

- ✔ know that organs work together in organ systems
- ✔ know the major organs of the digestive system

Key words

organ system, digestive system

Putting systems in place

Specialised cells are combined in tissues, and tissues work together in organs. But even organs do not work alone – several organs may be organised to work together to achieve the life processes of the organism. A group of organs working together is called an **organ system**. There are many organ systems in the human body, but perhaps the best known is the **digestive system**.

The organs of the digestive system

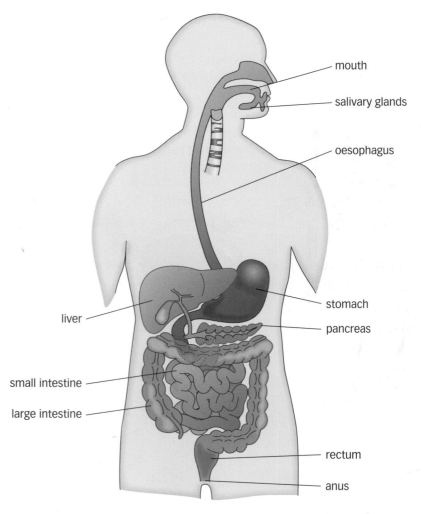

In the digestive system several organs work together to bring about digestion

Exam tip   AQA

- ✔ Make sure you can label the organs of the digestive system on a diagram.

A What is an organ system?

B Apart from the digestive system, name another organ system in the human body.

C Explain why the digestive system needs several different organs to carry out its function.

Division of labour

The function of the digestive system is the digestion and absorption of food. The different organs in the digestive system each have different functions in this overall task.

The digestive system allows the exchange of substances between the body and its environment. The body takes in food into the gut, and releases digestive juices to mix with the food. The resulting digested food particles and water pass through the walls of the digestive system. They are taken into the blood to be transported around the body.

Organs for releasing digestive juices

Some organs in the digestive system are involved in the production and release of digestive juices into the gut. These organs are glands. The pancreas and salivary glands are two major glands that carry out this function.

Digestive juices can have two functions:
- to lubricate the food
- to carry enzymes to aid digestion.

In addition, the juice released by the pancreas together with the bile from the liver also helps to neutralise acid formed in the stomach.

Organs for digestion

Digestion is the breakdown of large food molecules into smaller particles that can pass through the gut wall and be absorbed into the blood. This breakdown occurs inside the mouth, stomach, and small intestine. Each of these organs has glands that produce enzymes, and these are mixed with the food. The mixing is brought about by either the teeth and tongue in the mouth, or muscles in the wall of the stomach and small intestine. The digestive enzymes bring about the digestion of the food.

Organs for absorption

Absorption is the process by which the smaller digested food particles are taken into the blood. The particles pass through the gut wall and are taken into the blood from the gut environment.

Digested food particles are mainly absorbed in the small intestine. The lining of the small intestine has a very large surface area, making it efficient at absorption. The water in the undigested food is absorbed in the large intestine. At the end of the digestive process, the material that is left is faeces, which leave the body through the anus.

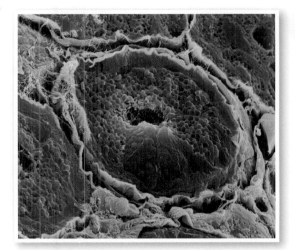

▲ A scanning electron micrograph of part of a salivary gland, where digestive juices are produced (× 1000)

Questions

1 Name two organs that produce digestive juices.

2 What substances are taken into the body from the external environment via the digestive system?   ↓ E

3 Explain why the digestive system is regarded as an organ system.

4 Name one other organ system in the human body involved in exchange, and state what it exchanges.   ↓ C

5 Explain how the small intestine is adapted for absorption.   ↓ A*

Learning objectives

After studying this topic, you should be able to:

- ✔ know that animal and plant cells are organised into tissues and organs
- ✔ know the main organs of the plant
- ✔ understand the distribution of tissues inside the plant

Key words

epidermal tissue, xylem, phloem, palisade mesophyll cells

Organising an organism

As with animals, plant cells are organised in a specific way within the plant:

- Groups of similar cells work together as a tissue.
- Groups of different tissues work together as an organ.
- All the organs build the whole organism.

In plants there are a number of different organs, each with a different function.

Plant organs

Organ	Function
Stem	Supports the plant. Transports substances through the plant.
Leaf	Produces food by photosynthesis.
Root	Anchors the plant. Takes up water and minerals from the soil.
Flower (this is an organ system consisting of three organs: the petal, the stamen, and the carpel)	Reproduction.

▲ The stem, root, and leaf are organs. The flower is an organ system.

A Name three plant organs.

B In an organ system, different organs work together. Why is the flower classed as an organ system?

Inside a plant

Inside a plant organ are tissues made up of similar cells working together:

- The outside of plant roots, stems, and leaves are covered in **epidermal tissue**. This protects the organs, although the root epidermis may get damaged by soil.
- The bulk of the stem and root is composed of packing cells. These cells are filled with watery fluid, which makes them firm so they can help support the plant.
- Two major tissues inside roots, stems, and leaves are **xylem** and **phloem**. These tissues are involved in transport, and they are found in the tube-like vascular bundles that run up through the roots, leaves, and stems. The cells of the xylem have thickened cell walls. These cells are strong and help support the plant.
- The leaf has many cells specialised for photosynthesis. These are the **palisade mesophyll cells**, located on the upper surface of the leaf. They contain lots of chloroplasts, and so can absorb sunlight energy for photosynthesis.

Transport in the vascular bundles

The vascular bundles form a continuous transport system from the roots, through the stem, and into the leaves.

There are two tissues inside the vascular bundles. Both are involved in the transport of water and dissolved substances through the plant:

- Xylem: these cells are dead and stacked on top of one another to form long hollow tube-like vessels. Xylem cells are involved in the transport of water and dissolved minerals from the roots to the shoots and leaves.
- Phloem: these cells are living and are also stacked on top of one another in tubes. They transport the food substances made in the leaf to all other parts of the plant.

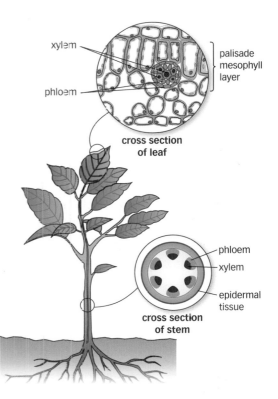

▲ A section through a leaf, showing the different tissues

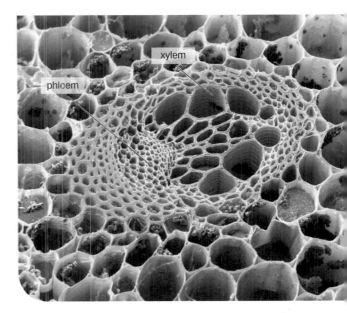

▲ A section through a buttercup stem to show the vascular bundles (×200)

Questions

1 What is the function of the mesophyll tissue?

2 Explain why it is important that xylem cells are hollow.

3 Explain how the plant supports itself.

E ↓ C ▼ A*

Learning objectives

After studying this topic, you should be able to:

✔ know that photosynthesis is the process by which plants make their own food

✔ appreciate the source of the raw materials for photosynthesis

✔ understand the fate of the products of photosynthesis

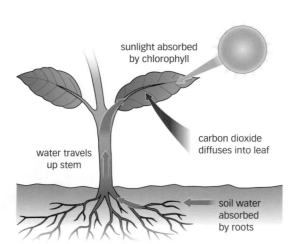

sunlight absorbed by chlorophyll

carbon dioxide diffuses into leaf

water travels up stem

soil water absorbed by roots

▲ In photosynthesis the plant uses sunlight energy to convert water and carbon dioxide into carbohydrates

Exam tip AQA

✔ If you are asked to write out an equation, make sure you know the difference between a word equation and a chemical equation. Learn the equations; it will gain you marks.

Feeding in plants

Plants do not take in ready-made food like animals do. They have to make their own food. To do this plants take in:

• carbon dioxide from the air
• water from the soil.

Some plants, algae, and seaweeds trap the Sun's energy in **chlorophyll**, in the chloroplasts in their cells. They use this energy to build up the carbon dioxide and water into carbohydrates and oxygen. This process is called **photosynthesis**.

Word equation for photosynthesis

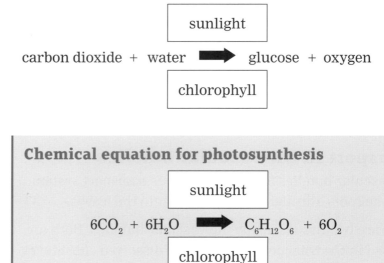

$$\text{carbon dioxide} + \text{water} \xrightarrow[\text{chlorophyll}]{\text{sunlight}} \text{glucose} + \text{oxygen}$$

Chemical equation for photosynthesis

$$6CO_2 + 6H_2O \xrightarrow[\text{chlorophyll}]{\text{sunlight}} C_6H_{12}O_6 + 6O_2$$

What does the plant make in photosynthesis?

You can see from the equations that there are two products of photosynthesis:

1. Glucose: this is food for plants, it is a carbohydrate. Some is used for respiration in the plant's cells. The rest can be stored in the plant.

2. Oxygen: this is a waste gas produced in photosynthesis. Some is used for respiration in the plant's cells. The rest is given off into the plant's surroundings. Without plants there would be no oxygen in the air for animals to breathe.

Converting glucose to other substances

The glucose produced in photosynthesis by plants and algae can be converted to other substances that the organisms need. For example, it may be used to make the sugar sucrose, found in sugar cane.

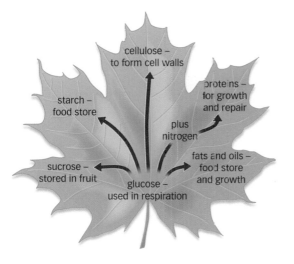

▲ Glucose from photosynthesis is converted to all the substances that a plant needs

If it is not used, the glucose can be changed into insoluble starch and stored until it is needed. Stored glucose can be used for respiration at night, when there is no sunlight and the plant is not making glucose by photosynthesis.

The glucose made in photosynthesis is converted to sucrose to be transported around the plant to parts that need it. Sucrose is good for transport because it dissolves in water and flows easily.

Plants are not made of sugars alone. The plant converts sugars to other substances such as cellulose, proteins, fats, and oils which it needs to grow and function. When producing proteins, plants do not just use sugars. They also use nitrite ions from the soil.

Storing glucose

Glucose is stored in the plant as starch. This has three advantages:
1. Starch can be converted back into glucose for respiration in plant cells.
2. Starch is insoluble and so will not dissolve in water and flow out of the cells where it is stored.
3. Starch does not affect the water concentration inside cells.

Key words

chlorophyll, photosynthesis

A What are the two raw materials a plant needs for photosynthesis?

B What else does a plant need in order for it to photosynthesise?

C Explain why humans could not survive without photosynthesis.

Questions

1 Where does the energy for photosynthesis come from? ↓ E

2 (a) What element is added to the glucose produced in photosynthesis to make proteins?
   (b) Where does this element come from? ↓ C

3 Explain why plant cells do not store carbohydrate as sugars. ↓ A*

Learning objectives

After studying this topic, you should be able to:

✔ know that the leaf is the site of photosynthesis

✔ appreciate the internal and external structure of the leaf

✔ understand the adaptations of the leaf for photosynthesis

Key words

leaf, palisade layer, stomata

Leaves

The main plant organs for making food are the leaves.

▲ A leaf   ▲ The external structure of a leaf

Inside the leaf

The **leaf** is made up of many specialised cells. Each type of cell has its own function. They work together, making the leaf well-adapted to carry out photosynthesis.

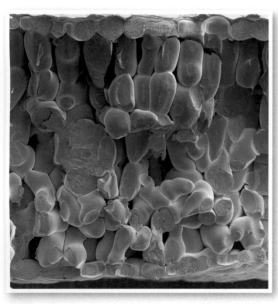

▲ Cross-section of a spinach leaf seen through a powerful electron microscope (× 250)

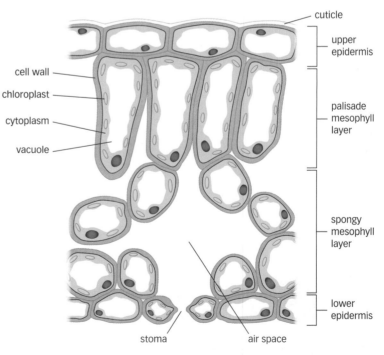

▲ The internal structure of a leaf

Exam tip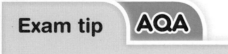

✔ You need to learn the structure of the leaf.

Top ten adaptations of the leaf for photosynthesis

✓ Many leaves are broad and flat, giving a large surface area to absorb as much light as possible.

✓ Leaves are thin, so that carbon dioxide has a short distance to travel to the mesophyll and palisade cells.

✓ The leaf cells contain chlorophyll within chloroplasts. This absorbs light energy for photosynthesis.

✓ The upper palisade layer, which receives the most light, contains the most chloroplasts.

✓ The cells of the **palisade layer** are neatly packed in rows, to fit more cells in.

✓ Veins carry water from the roots to the leaf cells, and carry glucose away.

✓ Veins support the leaf blade.

✓ There are plenty of **stomata**, pores in the lower epidermis, which allow carbon dioxide in and oxygen out.

✓ There are air spaces in the spongy mesophyll layer to allow carbon dioxide to diffuse from the stomata to the palisade cells.

✓ The air spaces inside the leaf give a large surface area to volume ratio. This allows maximum absorption of gases.

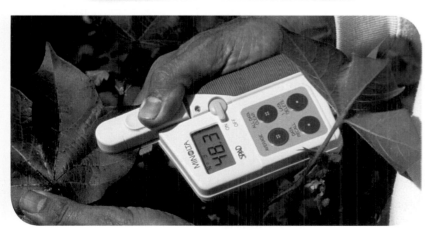

▲ This digital meter is being used to measure chlorophyll and photosynthesis in a cotton leaf

A On a plant, leaves are angled so plenty of sunlight reaches them. Explain why this is important to the plant.

B The leaf epidermis is transparent. Why is this an advantage to the leaf?

C What is the name for the pores in the leaf?

Questions

1 Name the layer in the leaf that carries out most photosynthesis. ↓ E

2 Which adaptations of the leaf allow it to trap as much sunlight as possible? ↓ C

3 Explain the advantages of the air spaces in the spongy mesophyll layer.

4 Leaves of plants that are often in bright sunlight tend to have more stomata. Explain what you think the effect of this will be. ↓ A*

Learning objectives

After studying this topic, you should be able to:

✔ know how some factors affect the rate of photosynthesis

✔ understand limiting factors

Key words

rate of photosynthesis, limiting factor

▲ The rate of photosynthesis in this greenhouse is increased using artificial lighting

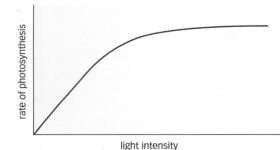

▲ Graph to show how the rate of photosynthesis changes as light intensity increases

The growing season

Plants do not grow at the same rate all year round. Most plants grow best in the spring and summer. This is when the conditions for growth are best. In spring and summer, the weather is usually warmer and there is more sunlight. These conditions are good for photosynthesis and therefore for growth, because the light energy is needed for photosynthesis, and the warmth speeds up the reactions of photosynthesis.

Increasing the rate of photosynthesis

The **rate of photosynthesis**, or how quickly the plant is photosynthesising, depends on several things. The following factors will speed up photosynthesis:

- more carbon dioxide
- more light
- a warm temperature.

People who grow plants commercially in a greenhouse try to make sure their plants have the best conditions. They use lighting systems which increase the hours of daylight available to plants, and they use heaters that burn gas, or other fuels, to add warmth and release carbon dioxide.

> A List three things that will increase the rate of photosynthesis.
>
> B Why do you think British woodland flowering plants such as bluebells flower in May?

Factors affecting the rate of photosynthesis

The rate of photosynthesis may be limited by the following factors.

Availability of light

Light provides the energy to drive photosynthesis. The more light there is, the faster the rate of photosynthesis. This is true provided that there is plenty of carbon dioxide, and the temperature is warm enough.

Amount of carbon dioxide

Carbon dioxide is one of the raw materials for photosynthesis. The more carbon dioxide there is available, the faster the rate of photosynthesis. (Again, this is only true if there is plenty of light and a suitable temperature.) Carbon dioxide is often the factor in shortest supply, so it is often the limiting factor for photosynthesis.

A suitable temperature

Temperature affects how quickly enzymes work. Enzymes make the reactions of photosynthesis happen. As the temperature rises, the rate of photosynthesis increases (providing there is plenty of carbon dioxide and light). However, if it becomes too hot, then the enzymes will be destroyed and photosynthesis stops.

Limiting factors

When a process is affected by several factors, the one that is at the lowest level will be the factor which limits the rate of reaction. This is called the **limiting factor**.

If the limiting factor is increased, then the rate of photosynthesis will increase until one of the other factors becomes limiting. For example, if photosynthesis is slow because there is not much light, then giving the plant more light will increase the rate of photosynthesis, up to a point. After that point, giving more light will not have any effect on photosynthesis, because light is no longer the limiting factor. The rate may now be limited by the level of carbon dioxide, for example.

▲ Warm, sunny conditions mean light and temperature are not limiting factors for photosynthesis

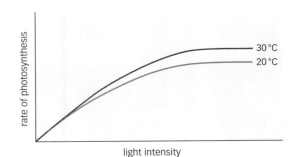

light intensity

▲ Light levels are limiting initially. The rate of photosynthesis then levels off. It increases at a higher temperature, so at higher light levels, temperature becomes the limiting factor.

Questions

1 If a plant receives more light, will its rate of photosynthesis increase or decrease?

2 Explain why burning a fuel in a greenhouse will increase the rate of photosynthesis.

3 Explain what a limiting factor is.

4 Explain in terms of limiting factors why gardeners do not need to mow lawns in winter.

Exam tip · AQA

✔ If you increase a limiting factor, then you will increase the rate. If you decrease the factor, then you decrease the rate. Remember to describe an increase or a decrease, rather than saying 'photosynthesis depends on the factor'.

Learning objectives

After studying this topic, you should be able to:

- ✔ understand that the factors needed for photosynthesis can be controlled
- ✔ appreciate that there are commercial benefits to controlling photosynthesis in greenhouses

▲ The Victorian Palm greenhouse at Kew Gardens in London

Did you know...?

The greenhouses at Kew, perhaps some of the best known in the world, originally had green glass.

Greenhouses

People have been growing plants in **greenhouses** for many years. The Victorians used greenhouses to grow rare tropical plants collected from their travels around the world. The great greenhouses at Kew Gardens date from that time. Greenhouses have also been used to grow food plants at unseasonal times of the year.

In modern greenhouses, growers use their understanding of photosynthesis to maximise the growth of plants. They artificially control the factors that limit the rate of photosynthesis.

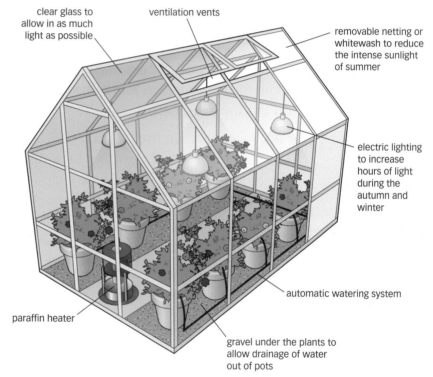

clear glass to allow in as much light as possible

ventilation vents

removable netting or whitewash to reduce the intense sunlight of summer

electric lighting to increase hours of light during the autumn and winter

automatic watering system

paraffin heater

gravel under the plants to allow drainage of water out of pots

▲ Modern greenhouses use automated systems to give the best conditions for photosynthesis

A Why were greenhouses first developed?

B Why do modern commercial greenhouses use automated systems?

Making the most of plants

Controlling light

Year-round light is provided by the Sun and by electric lighting systems. In the summer the sunlight can be too strong. Netting or whitewashing the windows can reduce the amount of light entering the greenhouse.

- **Advantage:** plants can receive light all year round, so they can be grown throughout the year.
- **Disadvantage:** cost of electric lighting.

Controlling temperature

The glass traps heat from the Sun inside the greenhouse, and shields the plants from the wind. This allows cultivation of plants that need warm temperatures, and allows plants to be grown out of season. If it gets too hot, automatic vents may open to allow hot air out of the greenhouse. In winter, additional heat can be supplied using heaters.

- **Advantage:** plants can be grown out of season, and more tropical plants can be grown in the UK.
- **Disadvantage:** cost of fuel for heaters.

Controlling carbon dioxide

Carbon dioxide is often the limiting factor for plant growth. Additional carbon dioxide can be added by burning fossil fuels such as paraffin in heaters.

- **Advantage:** additional carbon dioxide speeds up the rate of photosynthesis.
- **Disadvantage:** cost of fuel.

Controlling the water supply

Water is needed for photosynthesis. However, too much water can lead to plants rotting. Any watering system used by the gardener has to supply enough water, but not too much.

Automatic watering systems water at set times, or have sensors in the soil to detect how dry the soil might be. Water can be sprayed over the plants, or put directly into the gravel shelves beneath the plants. This will allow water to drain out of the plants so that they are not too wet, but will also act as a reservoir to allow water to soak up into the pots.

- **Advantage:** plants have a constant supply of water so they can photosynthesise.
- **Disadvantage:** cost of setting up the systems and of the electricity and water to run them.

Key words

greenhouse

Questions

1. Name four of the factors limiting photosynthesis that can be controlled in a greenhouse. ↓ E

2. Explain why out-of-season strawberries are more expensive than those grown outside in the summer. ↓ C

3. Explain why low-value crops like potatoes are not grown in greenhouses.

4. If a market gardener wanted to grow a crop of strawberries out of season, what factors would they have to consider to make a profit? ↓ A*

Learning objectives

After studying this topic, you should be able to:

✔ know about common sampling techniques

✔ understand how to use sampling techniques to collect good quality data

▲ Where do you start studying all the different species in this meadow?

▲ Students count how many organisms of a certain species are inside the quadrat. This gives a sample.

▲ Using a transect and quadrat to study the distribution of organisms across the field

There's a lot out there!

When biologists investigate where organisms live, they meet problems:

- There are very many different organisms.
- They seem to live all over the place.

It is difficult to make sense of the huge amounts of data.

To overcome these problems, biologists have devised a series of techniques to collect information about two things. First, they record the location of organisms of one species; this describes their **distribution**. Second, they record the number of organisms of a particular species in an area; this is the **population**.

Different populations live together in one area, and together they form a **community**. Biologists look for **relationships** between the organisms in a community by studying how their distributions overlap. They also study how factors in the environment affect their distributions.

To collect this information, biologists need techniques to:

- collect organisms
- count the number of organisms in each species
- record where the organisms are found
- collect accurate data
- collect the data fairly
- collect reliable data.

Biologists use a technique called **sampling**. This means counting a small number of the total population and working out the total from the sample.

Sampling techniques

1. Quadrats are square frames of a standard area. They are put on the ground to define an area. The numbers of organisms of particular species in the frame can then be counted.

2. Transect lines are tapes that are laid across an environment. You can count the organisms that touch the tape, such as plants on the ground, in order to study their distribution. Alternatively you can lay quadrats at regular intervals down the tape in order to record the distribution of the organisms inside.

When you have enough readings, it is possible to make estimates of the size of a population from your sample. You can also estimate the distribution of the population. But the numbers in the sample need to be accurate, reliable, and fairly collected.

Valid sampling

Being accurate

The apparatus should allow you to count a reasonably large number of the type of organism you are studying. For example, if you use a quadrat that is too small, then you will record fewer plants and animals. A small sample size is not very accurate and would not be reproducible.

Being reliable

Repeat readings make the data more reliable. If only one quadrat is recorded, then it might not represent the population accurately. The more quadrats you record, the more reliable your data will be.

Being fair

To be fair, all your readings should use the same equipment. They also need to be placed fairly. When recording distribution, quadrats can be placed at regular intervals along a transect. This avoids you choosing places that look promising, which would give biased readings. When estimating population size, quadrats should be placed randomly in an area, rather than choosing where to place them.

Factors affecting distribution

There are many factors that could affect where an organism lives. Collecting data about the distribution of particular species allows biologists to compare their distributions in relation to factors such as:

- temperature: for example, polar bears with adaptations to cold climates are found in the arctic
- nutrients: lions need gazelles to eat; both species are found together in the same area
- light: many plants live in sunny locations, or have adaptations to shady conditions
- availability of water: few species live in deserts
- availability of oxygen: plants and animals need oxygen to respire
- availability of carbon dioxide: plants grow well with plenty of carbon dioxide for photosynthesis.

Key words

distribution, population, community, relationship, sampling

A Describe what a quadrat is used for.

B How could a group of students record the distribution of limpets down a beach?

Exam tip AQA

- ✓ Useful memory aids for sampling are 'Accuracy using Apparatus' and 'Reliability needs Repeats'.

Questions

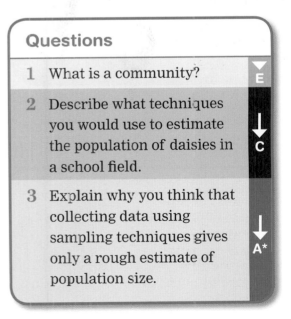

1 What is a community?

2 Describe what techniques you would use to estimate the population of daisies in a school field.

3 Explain why you think that collecting data using sampling techniques gives only a rough estimate of population size.

E ↓ C ↓ A*

Learning objectives

After studying this topic, you should be able to:

- ✔ know that environmental work can generate large amounts of data
- ✔ review some methods used to process data

Key words

mean, median, mode

▲ A researcher from Newcastle University monitoring water pollution. She will take several samples and analyse the data.

A Name three centralised values scientists generate from their data.

B Explain why centralised values are useful.

Numbers everywhere

Experiments generate lots of numbers. It can be hard to make sense of large collections of numerical data like this. Biologists process the data to try to look for meaningful patterns or relationships.

One common technique is to search for a centralised value that is typical of all the results and can be used to compare with other values. There are three ways of achieving this:

- The **mean** is the average value of the data. This is commonly used. For example, the researcher in the photograph might take several samples and quote the mean as the average level of pollution. A disadvantage is that the mean can be influenced by a rogue result which is very different from the other data.
- The **median** is the middle value of the data when arranged in rank order. So, the researcher could quote the middle value of all her levels of pollution. This is less affected by data points that are very high or very low compared with the others.
- The **mode** is the most common value of the data. Its advantage is that it is not affected by an extreme rogue result. The mode does not take account of the spread of the data.

Centralised values, or averages, like these are useful because they can give you a quick overview of what the data are showing. Biologists might use these centralised values and compare them with others, maybe from a different location, or to look for a relationship with other factors.

How scientists work
A case study: global warming

Scientists interpret data using centralised values. They analyse their data by looking for relationships. Any relationships that they find form their conclusions. However, other biologists might take the same sets of data and reach quite different conclusions from them.

One very important example is the evidence about environmental change.

Step 1: scientists make a hypothesis

People are concerned that global temperatures are rising, and that this is leading to habitat loss, such as the loss of the ice caps that are the habitat of the polar bear. Scientists think that rising levels of carbon dioxide in the atmosphere are causing this increase in temperature.

Step 2: scientists test their hypothesis

To test this idea, we need to collect data. Scientists take ice samples from the polar ice packs, formed over thousands of years. They record the carbon dioxide levels in air bubbles trapped in the ice. The structure of the ice also tells them about the temperature at the time it was formed. This gives a record of temperature and carbon dioxide levels over time.

▲ Scientists taking ice core samples

Step 3: analysing the data

From many samples, the scientists calculate the mean carbon dioxide level and temperature for each time period recorded. They then plot their data as a graph.

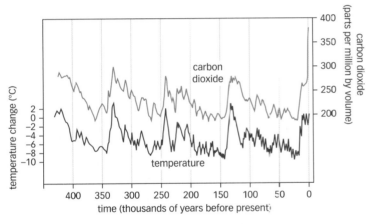

▲ Graph of carbon dioxide levels and temperature over the last 200 000 years

Step 4: interpreting the evidence

Some scientists are convinced that there is a direct relationship between the increase in global temperatures and the increase in the levels of carbon dioxide. They believe that human activity, including burning fossil fuels, is responsible for the increase in carbon dioxide. They also think that this climate change will have an impact on habitats and the distribution of many species.

Other scientists disagree with this interpretation. They feel that other factors, such as sun activity, have not been taken into account. At present the jury is still out!

Exam tip

✓ When calculating a mean average, only quote one more decimal place than that of your data.

Questions

1 Why do scientists take more than one ice core sample for each time period?

2 Why do scientists calculate mean values from all the samples at each time period?

⬇ E

3 Why do scientists plot temperature and carbon dioxide levels on the same graph?

⬇ C

4 Explain why it is important that other scientists test the findings of an experiment.

⬇ A*

Course catch-up

Revision checklist

- Living organisms are made of cells. Inside cells are parts, each of which has a particular function.
- There are similarities and differences between plant and animal cells.
- Fungal cells have a wall made from chitin.
- Bacterial cells have a wall, ribosomes, a membrane, and cytoplasm, but no nucleus.
- Cells have the same basic structure but they become specialised for different functions.
- Substances may move into and out of cells by diffusion. Diffusion is affected by distance, concentration gradient, surface area, and temperature.
- Groups of cells work together as tissues. Tissues work together as organs. Organs work together in systems in the organism.
- Different organs within each system have particular functions. The digestive system digests and absorbs food.
- In plants, roots, stems, leaves, and flowers are organs. Each has a specific function.
- Xylem and phloem are plant tissues. Xylem transports water and minerals; phloem transports dissolved food.
- Plants make their own food from carbon dioxide, water and light energy, by photosynthesis. The glucose they make can be converted to other sugars, starch, cellulose, fats and proteins (and nucleic acids). The waste product is oxygen.
- Leaves are organs adapted for photosynthesis. There are many of them, they are thin, have stomata for gaseous exchange, and have palisade cells with many chloroplasts.
- Rates of photosynthesis can be changed by levels of carbon dioxide, light intensity, and temperature.
- Humans can control the factors needed for photosynthesis and improve yields of commercially grown plants.
- Biologists use sampling techniques to investigate the biodiversity in different ecosystems, and the factors that affect animal and plant distribution. They analyse the data they collect.
- Scientists look for relationships from the analysed data. They then make a hypothesis and carry out experiments to test the hypothesis. Other scientists also repeat their investigations. If all their evidence supports the hypothesis, eventually it becomes accepted as a scientific theory.

blood

muscle

animal tissues

nerve

epithelium

if data supports the hypothesis it becomes a theory

do experiments to test hypothesis

other scientists also do these experiments

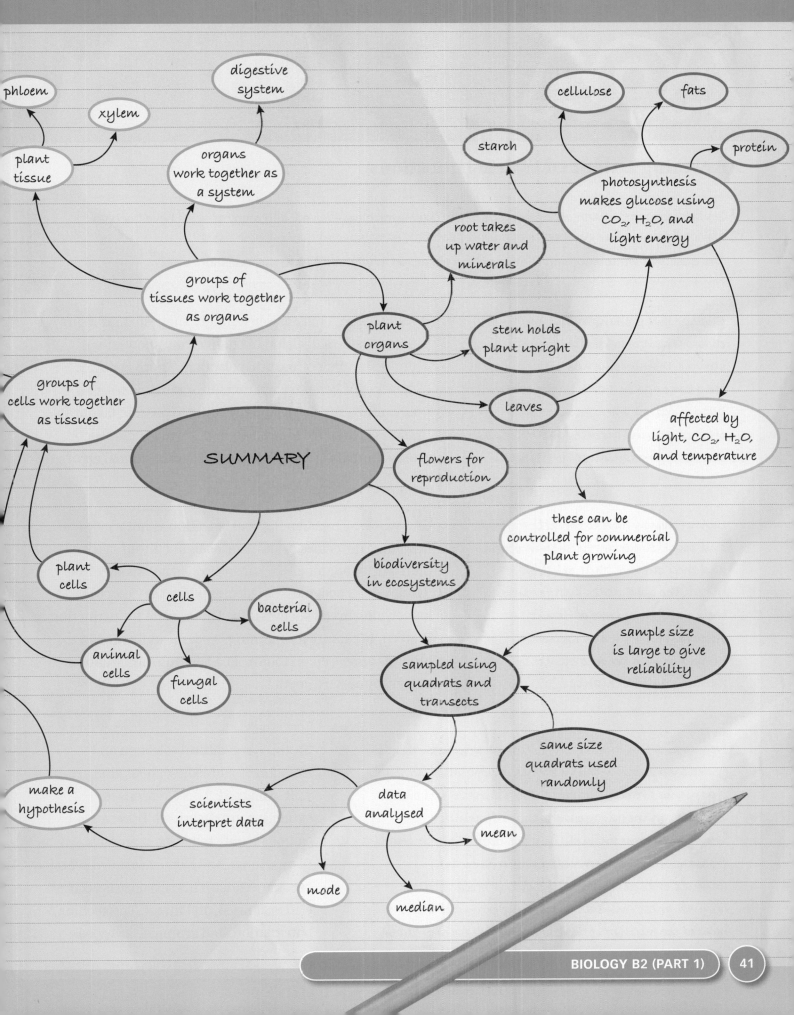

phloem

xylem

digestive system

cellulose

fats

plant tissue

organs work together as a system

starch

protein

photosynthesis makes glucose using CO_2, H_2O, and light energy

groups of tissues work together as organs

root takes up water and minerals

groups of cells work together as tissues

plant organs

stem holds plant upright

SUMMARY

leaves

affected by light, CO_2, H_2O, and temperature

flowers for reproduction

these can be controlled for commercial plant growing

plant cells

biodiversity in ecosystems

cells

bacterial cells

sample size is large to give reliability

animal cells

fungal cells

sampled using quadrats and transects

same size quadrats used randomly

make a hypothesis

scientists interpret data

data analysed

mean

mode

median

Answering Extended Writing questions

QUESTION

Since life began on Earth, many new species of organisms have evolved, and many species have become extinct. Explain how new species can evolve and how some species may become extinct.

The quality of written communication will be assessed in your answer to this question.

G–E

Living things change when the environment changes, so they can survive. they become new species. A long time ago on Earth lots of things became extinct when an asteroid crashed into us. Diseases and volcano's can also make animals go extinct.

Examiner: This answer wrongly suggests that living things make a decision to change, and shows no understanding of natural selection. There is no mention of genetic variation, or organisms with an advantage surviving and breeding to pass on favourable alleles. Three reasons for extinction are included, but only animals are mentioned. Two grammatical errors.

D–C

Living things have to compete for food and space. Some are fitter than others and they survive. Their children inherit the good genes and features. If their children become different enough they are a new species. Living things become extinct if humans kill too many.

Examiner: The term natural selection is not used. The candidate describes organisms as being 'fitter' rather than 'better adapted'. The answer does not specify that living things become new species when they can no longer interbreed successfully. One reason for extinction is given, but many living things became extinct before humans appeared on Earth. The spelling, punctuation, and grammar are good.

B–A*

New species can arise through natural selection. There's genetic variation within a species or population so some organisms are better adapted. These survive and breed and pass on the advantage to their offspring. Sometimes some animals in a population get separated from the others, by a mountain or a river. These animals change and then they can't breed with the original ones. Animals can become extinct if they are hunted too much or if there is a new disease or predator.

Examiner: Natural selection and isolation of a breeding population are well explained. Three reasons for extinction are described. However, the candidate only talks about animals. Plants, fungi, and bacteria also evolve or become extinct. This response is accurate, well organised, and fluent. The spelling, punctuation, and grammar are good.

Exam-style questions

1 The diagram shows a cell from the lung. Gases pass through this cell.

A01   **a**   Name parts A, B and C

A01   **b**   Which feature of this cell allows gases to pass through it?
    i   it has a large nucleus
    ii   it has many mitochondria
    iii   it is thin

A01   **c**   By what process does oxygen pass though this cell?
    i   osmosis
    ii   diffusion
    iii   respiration

A01   **d**   How is a bacterial cell different?
    i   it has no membrane
    ii   it has ribosomes
    iii   it has no nucleus

2 A student investigated how a leaf makes starch. Diagram 1 shows how he treated the leaf. Diagram 2 shows where starch was present after 8 hours.

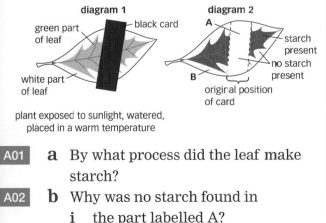

diagram 1
green part of leaf — black card
white part of leaf
plant exposed to sunlight, watered, placed in a warm temperature

diagram 2
A
starch present
no starch present
B
original position of card

A01   **a**   By what process did the leaf make starch?

A02   **b**   Why was no starch found in
    i   the part labelled A?
    ii   the part labelled B?

A02   **c**   Name the two independent variables in this investigation.

3 In an investigation potato chips were weighed before and after being placed in salt solutions for an hour.

salt concentration (M)	0.0	0.2	0.4	1.0	2.0
mass at start (g)	2.5	2.5	2.6	2.5	2.7
mass at end (g)	2.8	2.7	2.7	2.3	2.2
% change in mass	+12.0	+8.0		−8.0	−18.5

A03   **a**   Fill in the missing value.

A03   **b**   Why are the changes in mass expressed as a percentage change?

A01   **c**   By what process do cells in the chips gain or lose water?

A02   **d**   Name two factors that should be kept the same in this experiment to make it valid (fair).

A02   **e**   How could you find out the strength of salt solution that causes no change to the mass of the chip?

Extended Writing

4 Describe how you would find out which plants are present on a school playing field.   **A02**

5 Explain how leaves are well adapted for photosynthesis.   **A02**

6 Explain how commercial plant growers can manipulate environmental conditions to increase the yield of crops grown in greenhouses.   **A02**

G–E   D–C   B–A*

A01   Recall the science
A02   Apply your knowledge
A03   Evaluate and analyse the evidence

B2 Part 2

Genes and proteins, inheritance, gene technology, and speciation

Why study this unit?

You learn things in school to help you understand the world around you. Some of the main areas of research in biology today are genes, ageing, and regenerative medicine. In this unit you will learn how the genetic code governs the making of proteins in your cells, and why proteins are important. Enzymes are proteins, and are useful outside the body as well as inside. You may use washing powder containing enzymes, and enzymes are also used to help make some foods. You will find out why you need energy, and how your cells respire to release energy from the food you eat.

You will learn how characteristics governed by genes are inherited, and how genetic disorders occur. You will also learn about genetic engineering and gene therapy, cloning, embryo screening, genetic fingerprinting, and stem cell research. Our knowledge of genetics can also be used to help understand how new species develop.

You should remember

1 There are two types of cell division: mitosis for growth and asexual reproduction; and meiosis for sexual reproduction.

2 Cells need energy for their chemical reactions (metabolism) and for division.

3 Respiration releases energy from the food you eat.

4 Genes, on chromosomes, determine your characteristics.

5 The cell that you developed from was made from your mother's egg and your father's sperm, and each contained your parents' genes.

6 There is variation within and between species.

In 1996, a team of scientists at the Roslin Institute in Scotland, led by Professor Ian Wilmut, cloned a sheep using genetic material taken from a cell of an adult sheep. Dolly became the most photographed and most famous sheep in history. She was not the first mammal to be cloned, but was the first to be cloned from an adult cell that had undergone differentiation into an udder (mammary gland) cell.

Professor Wilmut is now director of the Centre for Regenerative Medicine at Edinburgh University, and his team works on stem cells and their potential use in treating brain and spinal injuries, and diabetes. The use of stem cells is not in itself controversial, but sourcing these cells from human embryos is. Scientists have recently found some stem cells deep in the dermis layer of the skin, which may prove a less controversial source of stem cells for research.

Learning objectives

After studying this topic, you should be able to:

- ✔ know that proteins are made of long chains of amino acids
- ✔ understand how the shape of a protein enables it to carry out its functions

Key words

proteins, antibodies, hormone, amino acids

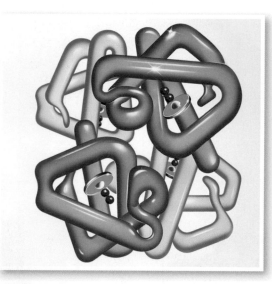

▲ The structure of haemoglobin. The pink and blue areas are proteins. The green areas are where the iron is. Each iron atom can hold two oxygen atoms, shown as red spheres.

Did you know...?

About 75% of your dry mass is protein. Because your cells contain a lot of water, this means that about 25% of the mass of each of your cells is protein.

Why are proteins important?

All the basic structural material of your body is made of **proteins**. Your skin, muscles, bones, cartilage, ligaments, and cell membranes all have a lot of protein in them.

In addition, you also have many other proteins in your cells. These other proteins do important jobs. They include:

- **antibodies**
- **hormones** and their receptors on membranes of target cells
- channels in cell membranes
- catalysts (enzymes that increase the rate of chemical reactions)
- structural components of tissues such as muscle.

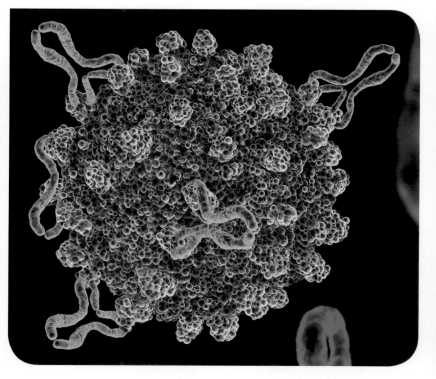

▲ Antibodies surrounding a virus particle in the blood. Antibodies are proteins. Each antibody fits on to the particular antigen, also made of protein, on the particular virus coat.

A Name five parts of your body that are made of protein.

B Name three other types of protein that help your cells to work properly.

What are proteins made of?

All protein molecules contain the elements:

- carbon
- hydrogen
- oxygen
- nitrogen.

In addition, some also contain sulfur.

Proteins are big molecules. They are made of smaller molecules, called **amino acids**. The amino acids are joined together to make long chains. These long chains then fold up into particular (specific) shapes. Each type of protein has a specific shape. Another molecule with a particular shape can fit into it.

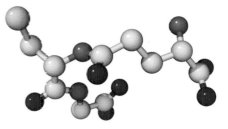

▲ Molecular model of an amino acid. The white balls represent carbon atoms; red represents oxygen; blue represents nitrogen; yellow represents sulfur. Hydrogen atoms are not shown.

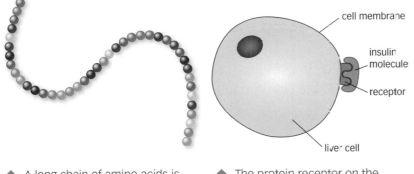

▲ A long chain of amino acids is called a protein

▲ The protein receptor on the surface membrane of a liver cell has a specific shape. The hormone insulin fits into it.

Labels: cell membrane, insulin molecule, receptor, liver cell

Exam tip AQA

✔ Remember that if two things are shaped so that one can fit into the other, they are not the same shape. Scientists say they are complementary in shape, but you can just say that one fits into the other. Think of a boiled egg fitting into an egg cup. They are not the same shape. You couldn't get one boiled egg to fit inside another boiled egg.

Questions

1 Name a hormone that is made of protein.

2 What elements are present in protein molecules?

3 What smaller molecules are joined together to make proteins?

4 Explain why the way each particular protein folds into a specific shape enables it to carry out its function (do its job).

5 A certain living organism contains 80% water. Some 75% of its dry mass is protein. How much of its total living mass is protein?

6 Fertiliser for plants contains nitrates, a form of nitrogen. Why do you think plants need nitrogen to grow?

7 Why do you think you need to eat a certain amount of protein each day?

8 When you eat proteins from meat, eggs, or soya beans, your body uses them to make the proteins in your cells and tissues. How do you think this happens?

↓ E

↓ C

↓ A*

Learning objectives

After studying this topic, you should be able to recall that enzymes:

✔ are proteins that catalyse chemical reactions in living cells

✔ have high specificity for their substrates

✔ work best at particular temperatures and pH

Key words

enzyme, catalyst, substrate, specific, optimum, denatured

A What is a catalyst?

B Name three types of chemical reaction that enzymes speed up in living organisms.

Enzymes are catalysts

Enzymes are **catalysts** because they speed up chemical reactions.

Most of these reactions, such as:

• photosynthesis
• respiration
• protein synthesis

take place inside living cells.

Enzymes can be used to catalyse the same type of reaction many times. This is like using one type of screwdriver to screw in many of the same type of screw, one at a time.

The shape of an enzyme is vital for its function

Enzymes, like all proteins, are folded into a particular shape. The shape of one particular area of the enzyme molecule, called the active site, is very important.

• The **substrate** molecules fit into the active site.
• This brings them together so they can form a bond.
• This makes a bigger molecule.

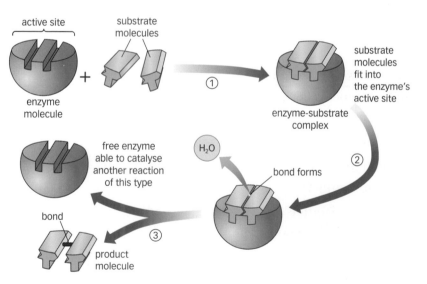

▲ The lock and key hypothesis is a hypothesis about how enzymes work. The two substrate molecules fit side by side into the enzyme's active site. A bond forms between them, and one large product molecule is formed.

In some cases (as shown on the left):

• A big substrate molecule fits into the active site.
• A bond breaks.
• Two smaller product molecules are made.

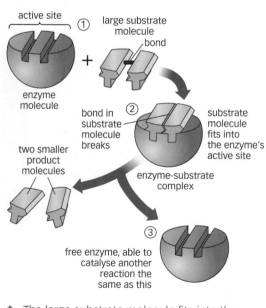

▲ The large substrate molecule fits into the enzyme's active site. A bond breaks and two product molecules are formed.

Enzymes have specificity for their substrate

- Only one particular type of substrate molecule can fit into an enzyme's active site. This is like the way only one type of key will fit into a particular lock.
- This means each enzyme is **specific** for its substrate molecules.

> C What are enzymes made of?
> D Why is the shape of the active site of an enzyme important?
> E How is each enzyme specific for a particular substrate?

What makes enzymes work best?

Each enzyme works best at

- a particular temperature, known as its **optimum** (best) temperature
- its optimum pH.

Low temperatures

The enzyme and substrate molecules have less energy. They do not move very fast so they do not collide (bump into each other) very often. The rate of reaction is low.

High temperatures

As the temperature increases, the enzyme and substrate molecules move more quickly and collide more often. This gives a faster rate of reaction.

However, if the temperature becomes *too* high then:

- The shape of the enzyme's active site changes.
- The substrate molecule cannot now fit into the active site.
- The rate of reaction slows and eventually stops.

When the shape of the enzyme has changed in this way, it cannot go back to its original shape. The change is irreversible. The enzyme is **denatured**.

pH

Each type of enzyme works at an optimum pH.

If the pH changes very much, then:

- The shape of the active site changes.
- The substrate molecules cannot fit into it.
- The enzyme has been denatured.

Did you know...?

Many enzymes in the body could work more quickly at temperatures above 37°C. However, if we kept our bodies hotter than this, many of our other proteins would be damaged. At 37°C our chemical reactions go on fast enough to sustain life. But some bacteria can live in very hot places, and their enzymes work well at 100°C.

Questions

1 State two conditions that enzymes need to work best. ↓ E

2 Explain why increasing the temperature from 10°C to 25°C makes the rate of an enzyme-controlled reaction increase.

3 Explain why if the temperature increased to 60°C, the rate of the enzyme-controlled reaction would slow down and eventually stop. ↓ C

4 As well as having enzymes inside your cells, you also have them in your blood. Your blood pH needs to be kept very close to 7.2. Why do you think this is? ↓ A*

Exam tip

- Always refer to the active site when you are explaining how an enzyme works.

Learning objectives

After studying this topic, you should be able to:

✔ understand that enzymes for digesting food work outside body cells

✔ know the roles of hydrochloric acid and bile in helping digestion

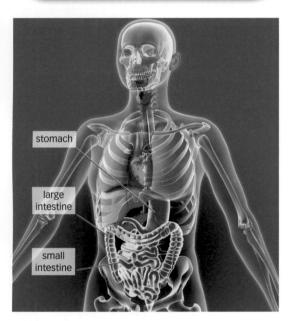

stomach

large intestine

small intestine

▲ Computer artwork showing the digestive tract (gut) inside the body

Did you know...?

You could digest your food without enzymes, but it would take several years to digest just one meal.

Exam tip 

✔ You should learn the information in the table on the next page. You need to know the names of these enzymes, where they are made and work, and what they do.

Why do you need to digest food?

It may seem a strange idea, but when food is in your gut it is still outside your body. Think of your body as being like a very long doughnut. The hole running through it is your gut. It is open to the outside at each end – the mouth and the anus.

Your gut or digestive tract runs from the mouth, via the gullet, stomach, small intestine, large intestine, and rectum to the anus.

The gut wall is part of your body of course, but the space in the middle of the gut is continuous with the outside of the body. So when the food you have chewed and swallowed is in your stomach and intestine, it is still outside.

You need to get it across the gut wall and into your bloodstream so it can go to your cells. The large food molecules have to be broken down into small molecules so they can diffuse across the gut wall and into the blood. This breaking down into smaller molecules is **digestion**.

> **A** Explain why you need to digest your food.

How is your food digested?

You have special enzymes that catalyse the breakdown (digestion) of large food molecules to smaller molecules. They are called digestive enzymes.

Digestive enzymes

- are made in specialised cells in glands and in the lining of the gut
- pass out of the cells where they are made and into the gut
- come into contact with food molecules and catalyse the breakdown of large food molecules into smaller molecules.

> **B** What is the substrate for the enzyme amylase?
> **C** What is the product when amylase catalyses the breakdown of its substrate?

Enzyme	Where it is made	What it does	Where it does it
Amylase	• In salivary glands • In the pancreas • In the lining of the small intestine	Catalyses the breakdown of starch molecules into sugar molecules.	In the mouth and in the small intestine.
Protease	• In the stomach • In the pancreas • In the lining of the small intestine	Catalyses the breakdown of protein molecules into amino acids.	In the stomach and small intestine.
Lipase	• In the pancreas • In the lining of the small intestine	Catalyses the breakdown of lipids (fat molecules) into fatty acid and glycerol molecules.	In the small intestine.

The best conditions

Enzymes need a particular pH to work best.

Some enzymes work well at acidic pH

Your stomach makes hydrochloric acid. This kills any bacteria that are in the food you eat. The proteases that work in your stomach can work well at low pH.

Some enzymes work best at alkaline pH

The enzymes that work in the small intestine work best in slightly alkaline conditions. When the food passes out of your stomach and into your small intestine, bile is released from the gall bladder. The bile enters the small intestine.

Bile
- is made in the liver
- is stored in the gall bladder
- neutralises the acid that was added to food in the stomach
- provides alkaline conditions in the small intestine for the enzymes there to work most effectively.

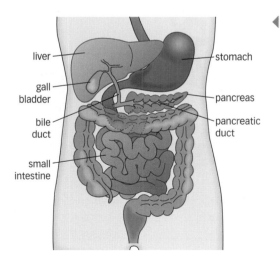

liver
gall bladder
bile duct
small intestine
stomach
pancreas
pancreatic duct

◄ In this diagram of the digestive tract you can see how the bile from the gall bladder passes down a tube and into the small intestine. You can also see how enzymes made in the pancreas can pass along the pancreatic duct and into the small intestine.

Questions

1 Name three places in the body where amylase is made.

2 What types of molecules are made when lipase catalyses the breakdown of fats?

3 What is the function of bile?

4 Where is bile made?

5 What is the function of hydrochloric acid in the stomach?

6 Does amylase work inside or outside body cells?

7 Protease enzymes that work in the stomach work well at very low pH. Protease enzymes that work in the small intestine work best in slightly alkaline conditions. Do you think they are exactly the same type of enzyme? Explain your answer.

↓E
↓C
↓A*

Key words

digestion, amylase, protease, lipase

17: Enzymes in the home – detergents

Learning objectives

After studying this topic, you should be able to:

✔ know how enzymes obtained from microorganisms are used in biological detergents

▲ Fermenters (special large containers) used for growing large numbers of bacteria. The bacteria inside make enzymes and pass them out of their cells. The enzymes can be collected from the liquid inside the fermenter.

▲ False colour scanning electron micrograph of granules of biological washing powder. Inside the opened granules you can see capsules of enzymes (×40).

What are detergents?

Detergents are cleaning agents. Washing powders contain detergents. Some washing powders also contain enzymes.

Obtaining enzymes from microorganisms

Some microorganisms make enzymes that pass out of their cells. If bacteria or fungi are grown in large vessels (called **fermenters**), scientists can collect the enzymes that they make and pass out of their cells.

> **A** What are detergents?
>
> **B** How are bacteria grown on a large scale?

Biological washing powders

Biological washing powders were first made on a large scale in 1965. They contain soap and have enzymes added to them. The enzymes can break down stains on textiles such as clothes. The stains may contain:

- protein – such as egg yolk or blood
- fats – such as grease or sweat.

The enzymes added to washing powders are proteases and lipases.

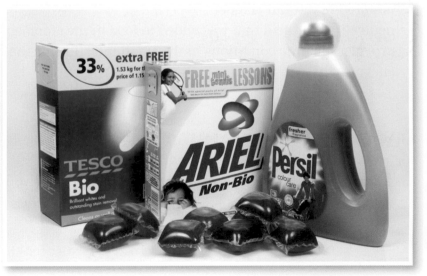

▲ Assorted washing powders. The biological washing powder on the left contains enzymes to break down fats and proteins in stains. Non-biological powders do not contain enzymes. Some people say they have had allergic reactions to the enzymes in biological washing powders. However, skin specialists in hospitals have carried out investigations. They have found that biological washing powders do not irritate skin.

C What sort of stains on clothes do protease enzymes break down?

D What sorts of stains on clothes do lipase enzymes break down?

Key words

detergent, fermenter

Washing powders are highly alkaline, so the enzymes added to them must be able to work well in solutions with high pH. They must also be able to work at a range of temperatures between 10 °C and 90 °C.

Some bacteria live in very hot places and have enzymes that work well at high temperatures. Some of the enzymes added to washing powders can be obtained from these bacteria.

Did you know...?

Some bacteria live near thermal vents on ocean beds at temperatures of 120 °C.

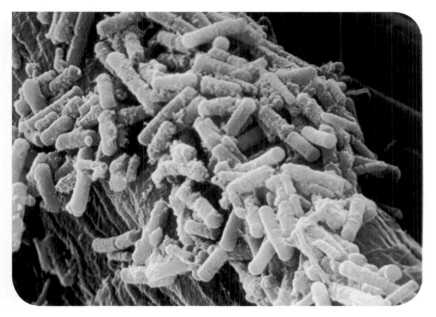

▲ These bacteria grow best between 55 °C and 75 °C and can survive at 90 °C. Their enzymes are not denatured at this high temperature (× 7200).

Low-temperature washing powders

Some biological washing powders are designed to be used at low temperatures. The enzymes added to them work well at low temperatures, so these washing powders remove stains better than other types of detergents at this temperature. Using cool water to wash clothes saves energy, and so saves money.

Some enzymes added to washing powders are denatured at high temperatures. These washing powders are designed to be used at temperatures up to 40 °C.

Questions

1 Why are enzymes for some biological washing powders obtained from bacteria that live in hot springs? ↓ E

2 What are the advantages of using biological washing powders at 20 °C?

3 What is the best pH for the enzymes added to washing powders? ↓ C

4 Explain what would happen to enzymes added to low-temperature washing powders if these powders were used in a hot wash at 80 °C. ↓ A*

Exam tip

✔ Remember that not all enzymes become denatured at temperatures above 40 °C.

Learning objectives

After studying this topic, you should be able to:

✔ understand why enzymes are used in industry

✔ consider some examples of using enzymes in industry

▲ The protein in baby food has been pre-digested with enzymes, so the baby can get all the amino acids from it

A Why do babies need to absorb amino acids from their food?

B Where are the enzymes used in industry obtained from?

Key words

carbohydrase, starch, glucose, fructose, isomerase

Did you know...?

Sugar is added even to savoury processed foods such as soup and baked beans.

Why use enzymes?

Many reactions used in industrial processes can take place without enzymes. However, the reaction mixture would have to be heated to a high temperature and kept at high pressure. This would use a lot of expensive energy. It would also need expensive equipment that could stand the high temperatures.

When enzymes are used to catalyse the reactions, the reactions can take place at lower temperatures and pressures. This saves money because the equipment used is not so expensive. Also, not so much fuel is needed.

Scientists can get enzymes from microorganisms. Some microorganisms make enzymes that pass out of their cells. Scientists grow lots of these bacteria in big vessels in laboratories, and then extract the enzymes.

Properties of enzymes used in industry

Industrial enzymes must

- have a long shelf life
- be able to stand fairly high temperatures
- have a wider than usual pH tolerance
- be able to work in the presence of chemicals that usually stop enzymes working.

How enzymes are used in industry

Proteases

Protease enzymes break down proteins. Some are used to pre-digest protein in baby foods. The enzymes break down the protein molecules into smaller molecules. Babies' stomachs are not strong enough to digest bigger protein molecules. Pre-digesting protein in this way makes sure that the baby can absorb amino acids from the baby food.

Carbohydrases

Carbohydrases are enzymes that break down **starch** into sugar (**glucose**) syrup.

Many processed foods have sugar added to them for flavour or to make the food taste sweet. Cane sugar (the sort you may put in your tea or coffee) is expensive. Food manufacturers use carbohydrase enzymes, such as amylase, from microorganisms. They mix the enzymes with cheap starch obtained from things like corn stalks.

◀ Corn stalks are a source of cheap starch

Exam tip

✔ Some exam questions are meant to be hard. You need to think about the question and make a sensible suggestion based on what you know – in other words, apply your knowledge to a new situation.

Glucose is added to some foods such as ice cream, and to some drinks. But it is not as sweet as cane sugar.

For some processed foods, a lot of glucose would need to be added. A type of sugar called **fructose** is much sweeter than glucose. Natural fructose is very expensive, but another type of enzyme from microorganisms, called an **isomerase**, can change glucose into cheap fructose.

Isomerase enzymes

An isomerase enzyme is mixed with the glucose syrup. This enzyme catalyses the changing of glucose molecules into fructose molecules. Both sugar molecules have the same number of atoms, but they are arranged slightly differently. Glucose and fructose molecules have slightly different shapes. Fructose tastes sweeter than glucose, so food manufacturers need to add less of it to their products. This is good for slimming foods.

▲ Slimming products

Disadvantages of using enzymes in industrial processes

Although enzymes from microorganisms are tougher than other enzymes, at very high temperatures some are denatured. Also, the product may be contaminated with some enzyme molecules. However, if the enzymes are enclosed in special little capsules, this protects them from the high temperatures and stops them contaminating the product.

Questions

1 Why do food manufacturers want to use fructose rather than glucose? ↓ E

2 What are the advantages of using enzymes obtained from microorganisms in chemical reactions used in industry?

3 Describe how carbohydrase enzymes are used to obtain glucose sugar. ↓ C

4 Describe how isomerase enzymes are used to turn glucose into fructose.

5 What are the disadvantages of using enzymes in industrial processes?

6 Fructose and glucose are both sugars. They have slightly differently shaped molecules. Why do you think they do not taste the same? (You have to think about this question and make a suggestion. Think about the shapes of protein molecules and taste receptors on your tongue.) ↓ A*

Learning objectives

After studying this topic, you should be able to:

✔ know that the energy needed for all life processes is provided by respiration

✔ understand that respiration can take place aerobically or anaerobically

▲ Buffalo (*Bison bison*) grazing on grass

A State three reasons why living things need energy.

B Name three types of large molecules that are made in living cells using energy.

▲ This grey wolf (*Canis lupus*) needs energy to run

What is energy?

Energy is the ability to do work. All matter has energy. There are different forms, such as kinetic (movement), potential (stored), heat, sound, electrical, and light energy. Each form of energy can be transferred into another form.

- Plants trap sunlight energy and use it to make large molecules – proteins, fats, and carbohydrates. These molecules contain stored energy.
- Animals get these molecules, containing stored energy, by eating plants or eating other animals that have eaten plants.

Why do living organisms need energy?

All life processes in all living organisms (including plants as well as animals) need energy. The energy may be used

- to build large molecules from smaller ones
- for muscle contraction in animals
- to control body temperature in mammals and birds.

Building large molecules from smaller ones

- Plants use sugars, nitrates, and other nutrients to make amino acids.
- Amino acids are joined together in long chains during protein synthesis. All living things need to make proteins such as enzymes and parts of their structure.
- Plants join sugar molecules together to make starch.
- Animals join sugar molecules together to make glycogen, which is similar to starch.
- Living organisms join fatty acids and glycerol together to make lipids (fats).

Muscle contraction

Animals need to move, to find food or a mate, or to escape from predators. Muscle contraction needs energy and causes movement.

Controlling body temperature

Some organisms cannot control their temperature very well. As the surrounding temperature changes, their temperature may also change. They control it by moving into the shade or into a warmer place. Snakes and lizards are very slow and sluggish in winter, or at night, when it is cold.

Birds and mammals can be active at night and during the winter. This is because a lot of the energy from the food they eat is released as heat energy. This keeps their body temperature steady regardless of the external temperature. However, it means that they need to eat more food than animals such as fish, snakes, and lizards.

How is energy released from food molecules?

Respiration in living cells releases energy from glucose molecules. You get glucose when you digest carbohydrates that you eat.

Respiration is a process that involves many chemical reactions, all controlled by particular enzymes.

Aerobic respiration

Aerobic respiration uses oxygen. It happens continuously in the cells of plants and animals.

Anaerobic respiration

Anaerobic respiration is a different type of respiration, that takes place without oxygen. This does *not* happen continuously in plant and animal cells. It happens when cells are not getting enough oxygen.

▲ Eagle owl (*Bubo bubo*) hunting at night

Did you know...?

The average temperatures of different species of birds and mammals vary.

Average body temperatures of some mammals and birds

Animal	Average body temperature (°C)
human	37.0
chimpanzee	37.0
dog	38.0
cat	39.0
rabbit	39.5
chicken	42.0
owl	38.5
eagle	48.0
penguin	38.0

Questions

1 Explain why animals need energy for movement.

2 What process in cells releases energy from food?

3 Explain why birds and mammals can be active at night when it is cold.

4 Explain the difference between aerobic and anaerobic respiration.

5 On a cold night in winter, a robin will lose a quarter of its body mass. Why do you think this is?

6 During the winter in the UK, many birds, such as swallows, cannot find enough food to eat to keep warm. How do you think they solve the problem?

7 During the winter in the UK, some mammals, such as hedgehogs, cannot find enough food to eat to keep warm. How do you think they solve the problem?

Key words

respiration, aerobic, anaerobic

Exam tip AQA

✓ Remember that plants respire all the time, just as animals do.

Learning objectives

After studying this topic, you should be able to:

- ✔ know the summary equation for aerobic respiration
- ✔ know that most of the reactions in aerobic respiration take place in mitochondria
- ✔ understand the changes that take place in the body during exercise

The equation for respiration

During aerobic respiration in living cells, there are chemical reactions which

- use glucose sugar and oxygen
- and release energy.

Although aerobic respiration involves a series of several chemical reactions, it can be summarised by the following equation:

glucose + oxygen → carbon dioxide + water (+ energy)

Mitochondria

Most of these chemical reactions take place inside **mitochondria**. These are sausage-shaped organelles (tiny organs) inside both plant and animal cells.

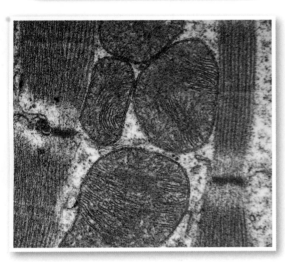

▲ Electron micrograph showing mitochondria, stained pink, in heart muscle (× 78 800)

> A   Write down a word equation for aerobic respiration.
>
> B   Where in your cells do most of the chemical reactions involved in aerobic respiration take place?

Energy and exercise

The energy released by respiration is used by the organism. One of the things it may be used for is muscle contraction.

When you exercise, your muscles need to contract more. They need more energy. They have to carry out more respiration to release the extra energy.

More respiration means the muscle cells need

- more glucose
- more oxygen.

To meet those needs

- your rate and depth of **breathing** goes up, which gets more oxygen into the lungs and into the blood
- your **heart rate** goes up, so more blood with oxygen is quickly delivered to the muscle cells, and this also delivers more glucose to muscle cells
- your muscle cells break down more of their stored **glycogen** into glucose.

▲ This runner needs energy released from respiration. The energy enables her muscles to contract.

How do your muscles store glycogen?

When you eat carbohydrates, your enzymes digest them to sugar.

The sugar passes across the gut wall into the bloodstream.

Some of that sugar goes from the blood into muscle cells. Here it is changed into glycogen (a big molecule similar to starch) and stored.

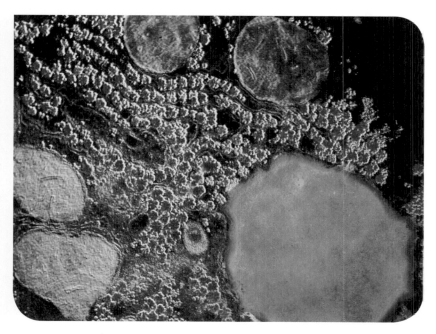

▲ False colour electron micrograph of stored glycogen (coloured pink) in a cell (× 9000)

When muscle cells respire more during exercise, as well as releasing more energy for their contraction they also release
- more heat
- more carbon dioxide.

Your body must not get too hot. The increased blood supply to and from your muscles carries away the extra heat to the skin. It also carries away the extra carbon dioxide to the lungs to be breathed out. Your heart rate and breathing rate stay high for a while after you finish exercising. This helps remove the extra carbon dioxide and heat.

Key words

mitochondria, breathing, heart rate, glycogen

Exam tip

✔ Do not confuse breathing with respiration. Respiration happens in cells and releases energy from food. Breathing is your chest movements that get air into and out of the lungs.

Did you know...?

Your brain detects the extra carbon dioxide in your blood. It sends signals to the heart and lungs to make heart rate and breathing rate increase.

Questions

1 Why do your muscle cells respire more when you run?

2 Explain why your breathing rate increases when you play a game such as football.

⬇ E

3 Explain why your heart rate increases when you ride a bicycle.

4 Why do muscle cells store glycogen?

⬇ C

5 Why do you think your blood carries heat to your skin?

⬇ A*

Learning objectives

After studying this topic, you should be able to:

✔ know that muscles use anaerobic respiration when they do not receive enough oxygen

✔ understand why muscles become fatigued after a long period of vigorous activity

A Write a word equation for anaerobic respiration.

B Where, in cells, does anaerobic respiration take place?

When your muscle cells do not receive enough oxygen

You have seen that during exercise your muscle cells need to respire more. Your breathing and heart rates increase to try to meet that need. However, sometimes they do not deliver enough extra oxygen to respiring muscle cells during exercise.

When this happens, your muscle cells use another type of respiration as well. They use anaerobic respiration. Anaerobic respiration happens in the cytoplasm of cells.

During anaerobic respiration, glucose is incompletely broken down to **lactic acid** instead of carbon dioxide. A much smaller amount of energy is released per molecule of glucose than in aerobic respiration. However, this incomplete breakdown happens quickly. So, many molecules of glucose can be broken down to meet the muscle's extra needs for a while.

You cannot use anaerobic respiration for long because the build-up of lactic acid is toxic. It causes muscles to become **fatigued**. They stop contracting efficiently. As your blood flows through your muscles it carries away the lactic acid. The blood takes the lactic acid to your liver, to be processed.

▲ Muscles involved during a golf swing use anaerobic respiration to release the energy needed

Did you know...?

Some kinds of exercise use anaerobic respiration because they happen quickly and there is no time for your heart and breathing rates to increase. A golf swing, tennis serve, a shotput, and a 100 m sprint all use anaerobic respiration to provide the energy for muscle contraction.

▲ This runner is suffering from a 'stitch', caused by lactic acid build-up

Key words

lactic acid, fatigue, **oxygen debt**

The oxygen debt

If you use anaerobic respiration to provide energy, then the muscle cells break down glucose to lactic acid. The lactic acid lowers the pH. This reduces the activity of the enzymes in muscles so they do not contract so efficiently. This is why you cannot exercise for long using anaerobic respiration.

When your heart rate and breathing rate have increased, you can use some of the extra oxygen to oxidise the lactic acid to carbon dioxide and water. This is called repaying the **oxygen debt**.

Questions

1 Which provides more energy for each molecule of glucose – anaerobic or aerobic respiration? ↓ E

2 What causes muscle fatigue? ↓ C

3 How does lactic acid affect the pH of your blood?

4 Why do you think you use anaerobic respiration if you run a 100 m sprint?

5 Why do you think an athlete's heart rate and breathing rate stay high for several minutes after running a 100 m sprint? ↓ A*

Exam tip

✔ Remember that aerobic means with oxygen. The prefix 'an' or 'a' means without. So anaerobic means 'without oxygen'.

Learning objectives

After studying this topic, you should be able to:

✔ know that mitosis is a type of cell division in body cells, producing two genetically identical cells

✔ know that organisms that reproduce asexually use mitosis

Key words

mitosis, chromosome, diploid, asexual reproduction, allele, growth

Exam tip AQA

✔ Mitosis replaces damaged cells and repairs tissue. It does not repair cells.

Why do body cells divide?

Body cells divide
- to replace worn-out cells
- to repair damaged tissues
- to grow by producing more cells.

Body cells divide by **mitosis**. Each cell produces two genetically identical daughter cells. This increases the total number of cells in a multicellular organism.

In the nucleus of most of your body cells you have two sets of **chromosomes**, arranged as 23 matching pairs of chromosomes. These cells are described as **diploid**.

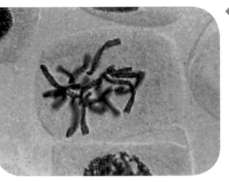

◀ This cell is about to divide. The chromosomes have coiled and become visible.

A Name a type of cell in your body that does not have any chromosomes in it.

Copying the cell's genetic material

Before a cell divides, its genetic material has to be copied. Each chromosome, made of one molecule of DNA (the genetic material), is copied. So, before a cell divides, each molecule of DNA copies itself. This is called DNA replication.

How DNA replicates

- DNA is a double-stranded molecule.
- The molecule 'unzips', forming two new strands.
- This exposes the DNA bases on each strand.
- Spare DNA bases in the nucleus line up against each separated strand of DNA.
- They only align next to their complementary DNA base, forming base pairs.
- One molecule of DNA has become two identical molecules.

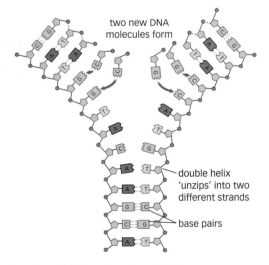

two new DNA molecules form

double helix 'unzips' into two different strands

base pairs

▲ How a DNA molecule replicates

How mitosis happens

- When each chromosome has made a copy of itself, these duplicated chromosomes line up across the centre of the cell.
- Then each 'double' chromosome splits into its two identical copies.
- Each copy moves to opposite poles (ends) of the cell.
- Two new nuclei form, each with a full set of chromosomes.
- The cell divides into two.
- Each cell is genetically identical to each other and to the parent cell.

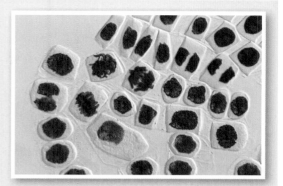

▲ Cells in the root tip of a hyacinth plant undergoing mitosis (× 180)

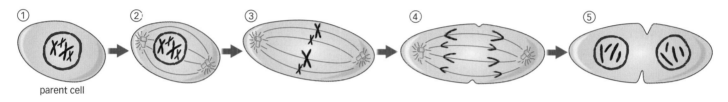

parent cell

▲ A cell dividing into two genetically identical cells by mitosis

Asexual reproduction

Some organisms can reproduce **asexually**. This type of reproduction also uses mitosis. The cells of the offspring produced by mitosis are genetically identical to the parent cells. They have the same **alleles** (versions of genes) as the parents.

Advantages of being multicellular

Early life forms on Earth were single-celled. There are still many simple single-celled organisms, such as the amoeba. Many organisms are now multicellular. Being multicellular means the organisms can be larger, can have different types of cells that do different types of jobs, and can be more complex.

Mitosis in mature organisms

In mature animals, cell division is mainly restricted to replacement of cells and repair of tissues. Mature animals do not continue to **grow**.

However, mature plants still have areas, such as root and shoot tips, where they can grow. The new cells made in these areas, by mitosis, can differentiate (become different and specialised) into many different types of plant cell.

Questions

1 Explain why a cell's genetic material has to be copied before it divides by mitosis. ↓ E
2 What is DNA replication?
3 Explain why mitosis is used for asexual division.
4 What are the advantages to an organism of being multicellular? ↓ C
5 How does the use of mitosis differ in mature plants and animals?
6 Explain how mitosis happens. ↓ A*

Learning objectives

After studying this topic, you should be able to:

- ✔ know that gametes are made by meiosis for sexual reproduction
- ✔ know that gametes are haploid and combine to give a diploid zygote
- ✔ understand that meiosis produces genetic variation

Key words

gametes, meiosis, haploid, zygote, fertilisation

> **A** What are gametes?
>
> **B** Where in the body are female gametes made? Where in the body are male gametes made?

Gametes

Gametes are sex cells. They are involved in sexual reproduction. Each gamete has only one set of chromosomes.

- Egg cells are made in the ovaries and sperm cells are made in the testes.
- Gametes are made by a special kind of cell division called **meiosis**.

How meiosis happens

- Just before the cell divides by meiosis, copies of the genetic information are made, just as they are before mitosis.
- So each chromosome has an exact copy of itself.
- However, in meiosis, the cell divides twice, forming four gametes.
- In the first division the chromosomes pair up in their matching pairs.
- They line up along the centre of the cell.

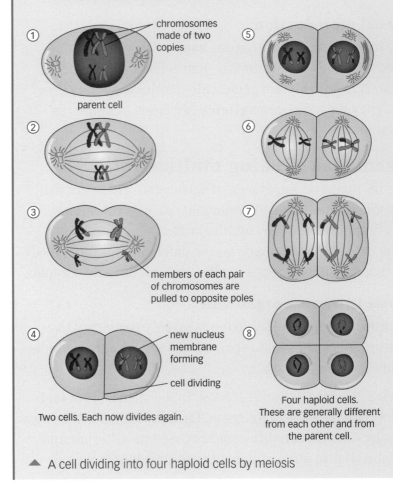

① chromosomes made of two copies

parent cell

②

③ members of each pair of chromosomes are pulled to opposite poles

④ new nucleus membrane forming

cell dividing

Two cells. Each now divides again.

⑤

⑥

⑦

⑧ Four haploid cells. These are generally different from each other and from the parent cell.

▲ A cell dividing into four haploid cells by meiosis

- The members of each pair split up and go to opposite poles (ends) of the cells.
- Now these two new cells each divide again.
- This time the double chromosomes split and go to opposite poles.
- Four cells, each having just one set of chromosomes, are made.

The cells made by meiosis are **haploid** gametes. They contain 23 chromosomes, *not* 23 pairs.

When two haploid gametes (an egg and a sperm) join, they produce a diploid cell called a **zygote**. This zygote will divide by mitosis into many cells and grow into a new individual:
- The joining of two gametes is called **fertilisation**.
- The combining of genetic material from two parents produces a unique individual.
- Half its chromosomes (and genes/alleles) have come from one parent and half from the other parent.
- It will have two sets of chromosomes.

A new individual will now develop from this cell, dividing many times by mitosis.

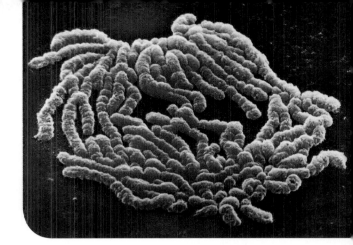

▲ Coloured electron micrograph (× 4100) showing chromosomes during meiosis

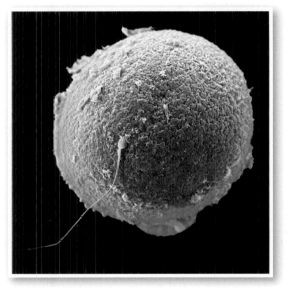

▲ Coloured electron micrograph showing a human sperm (coloured blue) penetrating a human egg (× 700)

Questions

1 How are gametes different from body cells?

2 Describe how male gametes are adapted to their function.

3 Describe how female gametes are adapted to their function.

↓ E

4 Why do sperm cells need a lot of energy?

5 By what process will the mitochondria in sperm provide energy?

6 What do the following terms mean?
(a) fertilisation, (b) haploid, (c) diploid, and (d) zygote.

↓ C

7 Explain why sexual reproduction produces genetically unique new individuals.

8 What type of cell division do you think causes the zygote to develop into an embryo?

↓ A*

Exam tip AQA

✓ Learn the spelling of meiosis. The word is similar to mitosis, so it has to be spelled correctly so that the examiner can be sure which type of cell division you are referring to.

Learning objectives

After studying this topic, you should be able to:

✔ be familiar with the principles used by Mendel in his investigations of monohybrid inheritance

▲ Gregor Mendel. Before he became a monk he was brought up on a farm where he gardened and kept bees.

▲ Pea plants (*Pisum sativum*)

Who was Mendel?

Gregor Mendel was born in 1822 in what is now the Czech Republic. He became a monk and studied maths and science at the University of Vienna. He used his monastery garden to study variation and inheritance in pea plants. He published a paper about his findings in 1866, but not many people read it and no one else seemed to be able to understand it. His work did not achieve the recognition it deserved in his lifetime, but was rediscovered about 40 years after it was published.

> **A** When was Gregor Mendel born?
>
> **B** Where was he brought up?
>
> **C** What did he study at Vienna University?

What did Mendel discover?

Before Mendel's plant-breeding experiments, people thought that characteristics were inherited by a blending mechanism. They thought, for example, that if a black dog and a white dog mated, then the puppies would be grey.

Mendel had access to the monastery garden and grew edible peas, *Pisum sativum*:

- He had several varieties of pea plants. He made sure, by sowing seeds from them and looking at the offspring, that each variety he used was true-breeding. For example, white-flowered plants always produced white-flowered offspring.
- He selected varieties with different characteristics.
- He cross-pollinated two different varieties. In one of his experiments he crossed tall-stemmed plants with short-stemmed plants.
- He took pollen (male gametes) from one variety of pea plant. He placed this on to the female part of the flowers of the other variety of pea plant.
- He then collected the seeds and grew them.
- He looked at the offspring (the F_1 **generation**) to see if they were tall- or short-stemmed.
- He saw they were all tall-stemmed. The short-stem characteristic seemed to have disappeared. He noted that they were not in between tall and short, so the blending idea did not seem to apply.

- He then allowed the plants of the F_1 generation to interbreed. He collected and grew their seeds.
- He looked at the offspring (F_2 generation) and saw that some were tall-stemmed and some short-stemmed.
- Then he did something very unusual for that time: he counted how many of each type there were.
- He found 787 tall plants and 277 short plants. He realised that there were three times as many tall plants as short plants.

He described the tall-stem characteristic as **dominant**. He described the short-stem characteristic as **recessive**.

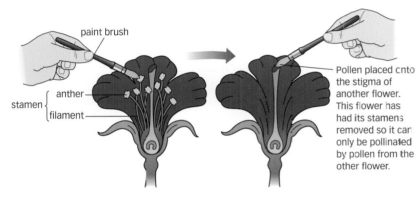

paint brush

stamen { anther / filament

Pollen placed onto the stigma of another flower. This flower has had its stamens removed so it can only be pollinated by pollen from the other flower.

▲ Cross-pollinating flowers – taking pollen from one flower and applying it to another flower to fertilise the egg cell and produce seeds. This technique allows researchers to carry out breeding experiments.

What did Mendel deduce?

Mendel said that inheritance was not by a blending mechanism but was due to **inheritance factors**. Factors for different characteristics are passed from parent to offspring separately.

He said:
- The cells in the pea plant each have two inheritance factors for every characteristic, such as stem length.
- Each pollen grain and each egg (female gamete) carries only one inheritance factor for a characteristic.
- Any pollen grain could fertilise any egg.
- If the offspring inherited one inheritance factor for tall stem and one inheritance factor for short stem, then it would show the dominant characteristic and be tall.
- If the offspring inherited two short-stem inheritance factors, then it would show the recessive characteristic; it would have a short stem.

Key words

F_1 generation, dominant, recessive, inheritance factor

Exam tip

✔ In his stem length experiment, Mendel was studying the inheritance of one characteristic. This is called monohybrid inheritance.

Questions

1 How did Mendel know if his peas were true-breeding?

2 Before Mendel did his inheritance experiments with peas, how did people at the time think that characteristics were inherited from parent to offspring?

↓ E

3 When he crossed true-breeding tall pea plants with true-breeding short pea plants, what were the offspring like?

4 When he allowed these offspring to interbreed, what were the F_2 offspring like?

↓ C

5 What conclusions did Mendel draw from the results of these experiments?

↓ A*

Learning objectives

After studying this topic, you should be able to:

✔ know how other scientists rediscovered Mendel's work

✔ understand that there are different forms of each gene, called alleles

✔ be able to interpret genetic diagrams of monohybrid inheritance

Key words

monohybrid inheritance, DNA (deoxyribonucleic acid), gene

A What is monohybrid inheritance?

B If a pea plant has inheritance factors **Tt**, is it tall- or short-stemmed?

Exam tip **AQA**

✔ Always follow the conventions for drawing genetic diagrams.

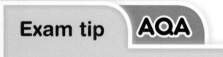

▲ Punnett square showing the monohybrid inheritance of crossing the F₁ generation

Genetic diagrams to explain Mendel's results

Genetic diagrams show how characteristics are inherited according to Mendel's laws. In each diagram we are looking at the inheritance of just one characteristic, such as the tall- or short-stemmed pea plants. The inheritance of one characteristic is described as **monohybrid inheritance**.

All of the first generation, the F_1 generation, were tall-stemmed. You can see that each plant had one dominant inherited factor and one recessive inherited factor. They all showed the dominant characteristic.

Mendel then interbred these **Tt** plants to form the F_2 generation.

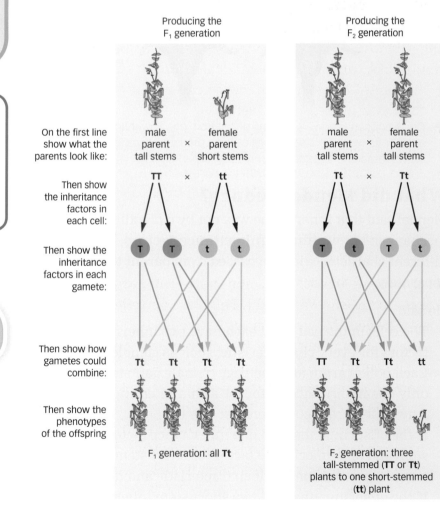

Instead of a genetic diagram, it may be clearer to use a Punnett square (see left) to show the inheritance.

You can see from this diagram that three out of the four possible combinations of factors produce tall-stemmed plants. One out of four possible combinations of factors produces short-stemmed plants. This gives a ratio of three tall-stemmed plants to one short-stemmed, which is what Mendel saw in his results.

Mendel did not get any prizes

By 1900 scientists had realised that the idea of blending inheritance was probably wrong. They were seeking a successful theory about separately inherited factors. Two biologists, independently of each other, duplicated the breeding experiments that Mendel had done about 40 years previously. When they read Mendel's account and conclusions they were able to understand their own findings.

Chromosomes, genes, and alleles

Scientists now know that the chemical that governs inheritance of characteristics is **DNA (deoxyribonucleic acid)**. DNA carries the genetic information:

- It is coiled into structures called chromosomes, which are in the nucleus of the cell.
- Each small section of DNA, called a **gene**, on a chromosome codes for a particular characteristic.
- Each pair of chromosomes contains the same genes, although each gene may have different forms called alleles.
- Each allele of a gene codes for the same characteristic, but for a slightly different version of it.

In the case of Mendel's tall- and short-stemmed pea plants, the characteristic was height. This was controlled by a single gene. The gene had two alleles:

- One, the dominant allele, had instructions for tall stems. This controls the development of the visible characteristic even if it is only present on one of the chromosomes.
- One, the recessive allele, had instructions for short stems. This controls the development of the visible characteristic only if the dominant allele is not present.

Questions

1 The diagram shows the results of crossing a pea plant having alleles **Tt** with a pea plant having alleles **tt**.

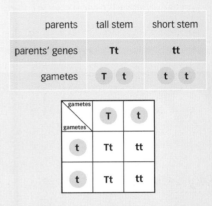

   (a) What do the offspring with **Tt** alleles look like?

   (b) What do the offspring with **tt** alleles look like?

   (c) If the original parent plants produced 300 seeds, how many would you expect to have the **Tt** alleles, and how many would you expect to have **tt** alleles?

2 Use a genetic diagram to predict the outcome of crossing a pea plant that has **TT** alleles with a pea plant that has **tt** alleles.

E

C

A*

▲ A pair of chromosomes and their genes. Many of the genes have different alleles on the two chromosomes.

Learning objectives

After studying this topic, you should be able to:

✔ know that one of the 23 pairs of human chromosomes carries the genes for sex (gender)

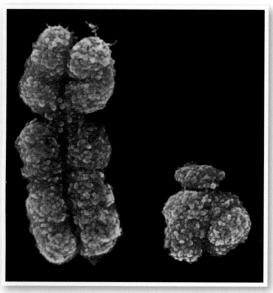

▲ XY chromosomes. The Y chromosome (blue) is much smaller than the X chromosome (pink). One small part of the X chromosome matches a small part of the Y chromosome, so that these two chromosomes can pair up for meiosis.

A Which of his sex chromosomes does a father pass to his son?

B Which of his sex chromosomes does a father pass to his daughter?

Did you know...?

Not all living organisms have the same mechanism as humans for determining sex. Most male mammals are XY, but male birds and butterflies are XX.

What makes us male or female?

You know that you have 23 pairs of chromosomes in the nucleus of all your body cells. One of these pairs of chromosomes determines sex. If, in this pair of **sex chromosomes**, you have two large X chromosomes, you are female. If you have one large X chromosome and a smaller Y chromosome, you are male.

Inheritance of sex

We can use a genetic diagram to show how sex (gender) can be inherited.

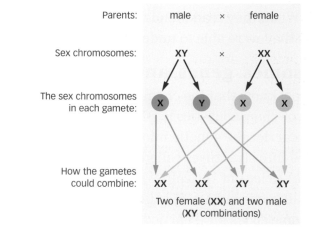

▲ There is an equal chance that each baby will be a girl or a boy

You can see that half the male gametes (sperm) have an X chromosome and half have a Y chromosome. However, all the female gametes (eggs) have an X chromosome.

Fertilisation is random, and either type of sperm could fertilise an egg. At each pregnancy there is a 50:50 chance of conceiving a girl or a boy. In a large population, there will be equal numbers of male and female offspring.

◀ Non-identical twins: boy and girl. Twins like these are rare. The mother produced two eggs at the same time. One was fertilised by a Y-bearing sperm and the other by an X-bearing sperm.

XX or XY?

The key gene, called SRY, which triggers a human embryo to develop into a male, is on the Y chromosome. This gene triggers the development of the testes. The absence of an SRY gene makes an embryo develop into a female.

Genetic diagrams

You may be asked to construct a genetic diagram for a monohybrid cross. You must follow the conventions:

- Show the characteristic of the parents.
- Show the alleles present in the parents' cells.
- Use upper case letters to represent a dominant allele.
- Use the lower case version of the same letter to show a recessive allele.
- Put gametes in circles.
- Show all the different possible combinations of alleles at fertilisation.
- Put an 'x' to show a cross (mating).

Here is a genetic diagram to explain a monohybrid cross between a father who has two alleles for free earlobes and a mother who has two alleles for attached earlobes.

Parents' characteristics: free earlobes × attached earlobes

Parents' alleles: (EE) × (ee)

Gametes: all (E) × all (e)

Offspring's alleles: all (Ee)

Offspring's characteristics: all have free earlobes but have one dominant and one recessive allele

Always make your upper case letters really big and the lower case letters small. And if you are told to use certain letters in your answer, then use them.

Key words

sex chromosomes

Questions

1 How many pairs of chromosomes do you have, in each cell nucleus, that do not play a part in determining your sex? ↓E

2 Explain how sex is determined in humans.

3 A couple have three children – all girls. What are the chances of them having a boy at the next pregnancy? Show how you worked out your answer. ↓C

4 A mother with blue eyes (recessive) and a father with brown eyes (dominant) have three children. Two of them have brown eyes and one has blue eyes. Construct a genetic diagram to explain how these parents can produce these offspring.

5 Some humans have XXY chromosomes in their cells. Do you think they will be male or female? Explain your answer.

6 Some genetic diseases, such as red–green colour blindness, are due to a faulty recessive allele on the X chromosome. Why do you think this disorder is more common in males than in females? ↓A*

Learning objectives

After studying this topic, you should be able to:

- ✔ know that chromosomes are made of DNA
- ✔ know that a gene is a small section of DNA
- ✔ understand that each gene codes for a particular protein

Key words

DNA bases

DNA

DNA stands for deoxyribonucleic acid. This chemical carries coded genetic information. Each molecule of DNA has a double helix structure. In the nucleus of all your cells you have 46 large molecules of DNA. Each of your chromosomes is one molecule of DNA.

You will remember from studying mitosis and meiosis that before a cell divides, all the molecules of DNA make copies of themselves. Then they condense into tightly coiled chromosomes so that they can move from the centre of the cell to the ends of the cell as the cell divides. When the cell has finished dividing into new cells, these chromosomes unravel. In that state they can govern the making of particular proteins.

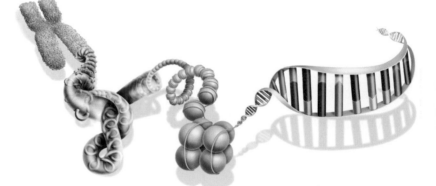

▲ How DNA condenses and coils into a chromosome. This DNA molecule has made a copy of itself so that when it forms a chromosome, the chromosome consists of the original and an identical copy. They are joined at a region near the middle. This gives the classic shape of visible chromosomes. It is only when chromosomes are coiled up like this that they take up stains and you can see them under a light microscope.

Genes

Each chromosome is one large molecule of DNA. Within each DNA molecule, shorter sections of DNA form specific genes.

Each gene has coded genetic information. It codes for a particular combination of amino acids that makes a specific protein.

Did you know...?

Scientists are now finding that some genes code for more than one protein.

A What does DNA stand for?

B How many molecules of DNA are there in a chromosome?

How genes code for proteins

Each section of DNA has a sequence of **DNA bases** in it. There are four bases, A, T, G, and C. You do not need to know their names.

1. These bases form a code. They are 'arranged' in groups of threes, or triplets.
2. Each triplet specifies a particular amino acid. So ATC will specify a different amino acid from ACT.
3. The sequence of base triplets on a section of DNA specifies the sequence of amino acids in a protein.
4. In turn, the sequence of amino acids in the protein governs how the protein will fold up into a particular shape.
5. Each different type of protein has a specific shape and can fit another molecule. This is how enzymes each fit just their own specific substrate molecule.

Exam tip AQA

✔ Remember that genes are on chromosomes. Each gene is a small section of DNA. Each chromosome is one huge molecule of DNA.

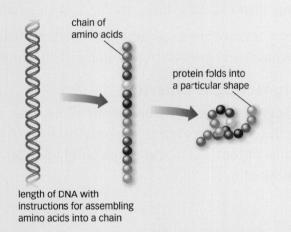

chain of amino acids

protein folds into a particular shape

length of DNA with instructions for assembling amino acids into a chain

▲ Simplified diagram to show how the coded information in a gene determines the shape and the function of a protein

All proteins have a specific shape, and this allows them to carry out their function. The characteristics that you inherit involve proteins. These characteristics may rely on the help of enzymes and hormones to develop, or they may directly involve a protein.

You have channels in your cell membranes that allow certain chemicals, such as chloride ions, into and out of the cells. If you have cells with faulty chloride ion channels in the membranes, you will suffer from cystic fibrosis (see spread B2.28).

Questions

1 Why does DNA condense and coil into chromosomes before a cell divides? ↓ E

2 Why does your DNA have to be unravelled when the cell is not dividing? ↓ C

3 Humans have about 20 000 genes. The two members of each pair of chromosomes have the same genes on them. So, on average, about how many genes do you think there are on each of your chromosomes? ↓ A*

4 Explain how different genes code for specific proteins.

Learning objectives

After studying this topic, you should be able to:

✔ know that some genetic disorders are inherited

✔ consider two genetic disorders – polydactyly and cystic fibrosis

Key words

digit, mucus, tract

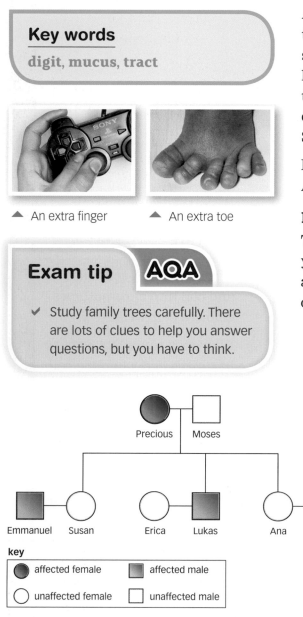

▲ An extra finger   ▲ An extra toe

Exam tip   AQA

✔ Study family trees carefully. There are lots of clues to help you answer questions, but you have to think.

Gene and chromosome defects

There are about 5000 genetic disorders caused by a defect in a single gene. There are also genetic disorders caused by whole chromosomes being abnormal. You need to consider just two examples of a genetic disorder.

Polydactyly

This simply means having more **digits** (fingers or toes) than usual (*poly* means many).

About 1 baby in every 500 births has one extra digit. This type of polydactyly is also called hexadactyly (*hexa* means six). The extra digit is usually on the little-finger side of the hand, or little-toe side of the foot. However, it may be on the thumb side as in the picture on the left. The extra digit causes no harm, but it is usually removed surgically at birth. Sometimes seven or nine digits are formed on a hand or foot.

Polydactyly is more common amongst black African and African American children than among white children.

How polydactyly is inherited

The condition is caused by a dominant allele. This means you have the condition even if you only have one faulty allele. If one parent has it, then each child has a 50% chance of inheriting it.

Precious   Moses

Emmanuel   Susan   Erica   Lukas   Ana   Syed

key

⬤ affected female   ⬛ affected male

◯ unaffected female   ▢ unaffected male

▲ A family tree showing incidence of polydactyly in a family

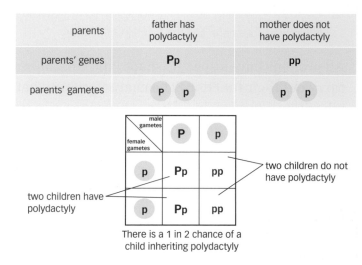

parents	father has polydactyly	mother does not have polydactyly
parents' genes	**Pp**	**pp**
parents' gametes	P p	p p

two children have polydactyly

two children do not have polydactyly

There is a 1 in 2 chance of a child inheriting polydactyly

▲ A genetic diagram showing how polydactyly is inherited

Cystic fibrosis

Cystic fibrosis (CF) is a disorder of cell membranes. The membranes of cells lining your airways and pancreas have faulty chloride ion channels. As a result you will have thick sticky **mucus** in these tubes or **tracts**. You get scarring (fibrosis) and cysts (cystic) in the pancreas. Other symptoms are

- frequent lung infections and difficulty breathing
- failure to thrive, as children cannot properly digest food because enzymes from the pancreas do not pass into the gut
- it shortens life, although antibiotics, physiotherapy, and lung transplants have extended life expectation. Many people with CF are now in their thirties and expect to live longer
- men and women may be infertile. Men can make sperm but do not have a sperm duct. Women have thick mucus that blocks their cervix.

How cystic fibrosis is inherited

CF is caused by a faulty, recessive allele of the gene for the cell-membrane-channel protein. A person with the disorder has to inherit two copies of the recessive allele – one from each parent. The parents may not have the disease. They may be symptomless carriers. They will each have one recessive allele and one normal, dominant allele. So they have enough functioning cell-membrane channels.
At each pregnancy for two carrier parents, there is a 25% chance that the child will have CF.

Questions

1. Look at the family tree for polydactyly. What are the chances that:

   (a) Syed and Ana will have a child with an extra digit?

   (b) Erica and Lukas will have a child with an extra digit? In each case explain how you arrived at your answer.

   ↓ E

2. Look at the CF family tree.

   (a) What are the chances of Julia and Michael having a child with CF?

   ↓ C

   (b) Joan and David are both carriers of CF. How can you tell this from the family tree?

   (c) If Joan and David have another child, what are the chances that it will have CF? In each case, show how you work out your answer.

   ↓ A*

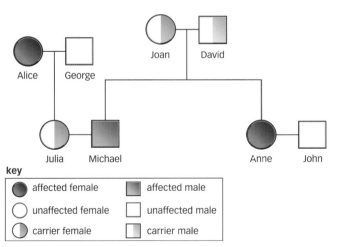

Alice **George** **Joan** **David**

Julia **Michael** **Anne** **John**

key
- ● affected female
- ■ affected male
- ○ unaffected female
- □ unaffected male
- ◑ carrier female
- ◧ carrier male

▲ A family tree showing incidence of CF within a family

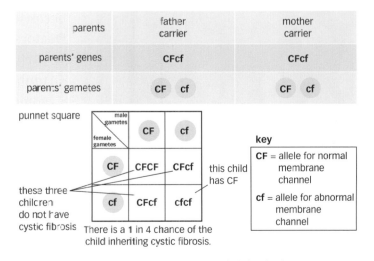

parents	father carrier	mother carrier
parents' genes	CFcf	CFcf
parents' gametes	CF cf	CF cf

punnet square

female gametes \ male gametes	CF	cf
CF	CFCF	CFcf
cf	CFcf	cfcf

this child has CF

these three children do not have cystic fibrosis

There is a 1 in 4 chance of the child inheriting cystic fibrosis.

key
CF = allele for normal membrane channel
cf = allele for abnormal membrane channel

▲ A genetic diagram showing how CF is inherited

Learning objectives

After studying this topic, you should be able to:

- ✔ know that embryos can be screened for alleles that cause genetic disorders
- ✔ understand that stem cells from embryos can be used for medical research
- ✔ explain how DNA fingerprinting can be used to identify individuals

Key words

stem cell, DNA fingerprinting

A Who might use embryo screening during pregnancy?

B List two advantages of using embryo screening.

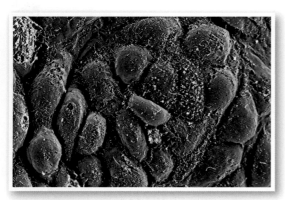

▲ Coloured scanning electron micrograph of human embryonic stem cells (× 1500). These were grown at the Centre for Life, Newcastle upon Tyne. Scientists can make them develop into any kind of human cell (and we have 220 different cell types) by giving them an appropriate chemical signal.

Embryo screening

If both parents know they are carriers of a serious genetic disorder, such as cystic fibrosis (CF), they may decide to have children in the following way:

- undergo in vitro fertilisation (IVF) to produce embryos
- have the embryos tested to see if they have CF (if they have two copies of the abnormal allele)
- only implant embryos that are free of CF.

There are several advantages of this:

- Their children will not have CF. Although CF can be treated, people with it have many health problems and may live shorter lives.
- Their children will not have the CF allele so will not be able to pass on the disorder to their own children.
- Although IVF is expensive, in the long term, money will be saved by the NHS as the children born will not have to be treated for CF. Also, they will not need a heart–lung transplant later in life.

However, as a result:

- Some embryos are formed that do not get the chance to develop into people.
- Some people who have a genetic disorder think this procedure is discriminating against them.

Stem cells

Stem cells are useful in medical research and treatments because they can develop into many different kinds of cell. Doctors can get stem cells from

- early embryos
- umbilical cord blood cells
- some types of adult tissues, such as bone marrow
- and more recently from amniotic fluid.

Using stem cells

Early embryo cells are the most useful because they still are able to differentiate into any type of cell. In mature animals, the ability of cells to differentiate is restricted to repair and replacement.

Medical research is developing ways of using stem cells to

- treat people with Parkinson's disease
- repair spinal-cord injuries
- grow tissues or organs for transplanting
- treat people with type 1 diabetes.

Scientists are allowed to keep spare embryos created by IVF alive for up to 14 days. Single cells taken from them are used as stem cells, to develop into different types of cells, such as nerve cells, blood cells, and muscle cells.

This raises ethical issues, as the spare embryos are not allowed to develop into people. However, even without stem-cell research these embryos would still be discarded.

DNA fingerprinting

In 1984 Professor Sir Alec Jeffreys, at the University of Leicester, discovered that certain parts of your DNA (not the genes, but other bits called VNTRs) vary a lot between people. However, they are similar within families. He used this knowledge to make **DNA fingerprints** of people.

Uses of DNA fingerprinting

The technique was used in a famous court case. A mother wanted to bring her child to the UK, but needed to prove that the child was hers. Alec Jeffreys did DNA fingerprints and proved that the woman was the mother of the girl.

Leicestershire Police heard about the technique and used it in 1986. It showed that a person who had confessed to a murder had not done it. Using DNA fingerprinting, the force caught the real murderer, Colin Pitchfork. He was the first criminal convicted in the UK using DNA evidence.

DNA fingerprinting is used to establish

- paternity and maternity
- criminal convictions for rape and murder
- whether a person is innocent of a crime after they were convicted
- family relationships.

The remains of some bones found in Russia were shown to be those of the Russian royal family, executed in 1917.

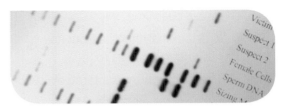

Did you know...?

Scientists can now take one cell from an eight-cell embryo and use that cell to grow more stem cells. The remaining seven-cell embryo can develop normally into a baby.

▲ DNA fingerprint results from a rape investigation. Suspect 1's DNA matches that of the sperm found in the victim.

Exam tip AQA

✓ When you answer questions like 2–5 below, you must be objective. Do not rant about your own personal opinions, but put forward reasons for and reasons against.

Questions

1 What are stem cells?

2 Think of reasons for and reasons against the following, writing down your ideas:

  (a) Having a national database with a sample of everyone's DNA on it.

  (b) Using embryo stem cells for medical research.

  (c) Embryo screening.

3 What sort of division do you think is used when a stem cell divides?

Learning objectives

After studying this topic, you should be able to:

✔ know that organisms have changed over time

✔ appreciate that there is evidence for evolution, such as fossils

✔ know how fossils were formed

✔ evaluate fossil evidence

Key words

evolution, fossil

Exam tip **AQA**

✔ Remember that fossils give good evidence for evolution, but they do not give a complete picture. You may be asked to evaluate how good the evidence is.

A What is a fossil?

B What are the four main types of fossil?

Evidence for evolution

The theory of **evolution** says that all the organisms around today have developed from previous life forms. According to the theory, organisms change gradually over time. Any proposed theory requires evidence to support it before it can be accepted.

Any scientific observation that supports the idea of evolution is part of the body of evidence for evolution. There are several types of evidence, but perhaps the most significant is the fossil record.

Fossils

Fossils are the preserved remains of living things from many years ago. There are many ways that fossils have been preserved.

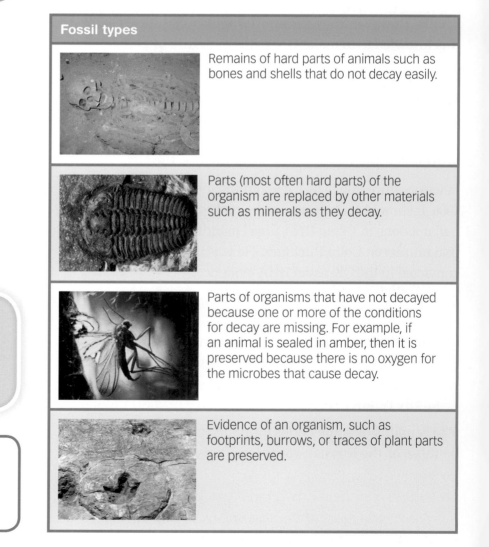

Fossil types	
	Remains of hard parts of animals such as bones and shells that do not decay easily.
	Parts (most often hard parts) of the organism are replaced by other materials such as minerals as they decay.
	Parts of organisms that have not decayed because one or more of the conditions for decay are missing. For example, if an animal is sealed in amber, then it is preserved because there is no oxygen for the microbes that cause decay.
	Evidence of an organism, such as footprints, burrows, or traces of plant parts are preserved.

How are fossils formed?

Perhaps the best known of the fossil types are those found in rock, formed as the hard parts of the organism are replaced by minerals. These fossils are formed like this:

- The plant or animal dies, and falls into soft mud or silt, often at the bottom of a lake or the sea.
- The body becomes covered in silt or mud.
- This gradually turns to rock, encasing the dead body.
- Over millions of years the hard parts of the body of the plant (leaves) or animal (bones and shells) are replaced by minerals. Soft parts of bodies do not always fossilise well, as they decay quickly.
- If earth movements make the land rise then the remains may become exposed at the surface.

Studying fossils

Biologists and palaeontologists study fossils, date them, and put them into date order. This can reveal the gradual change of one type of plant or animal into another over time. This can give evidence for the steps in evolution.

One of the best studied fossil records is that of the horse. From the record it can be seen that:

- The horse has grown, originally being the size of a dog.
- It now stands on long legs with only one digit.
- Its teeth are now adapted for eating grass.

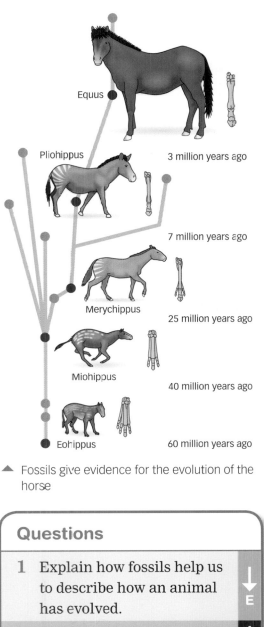

Equus

Pliohippus — 3 million years ago

7 million years ago

Merychippus — 25 million years ago

Miohippus — 40 million years ago

Eohippus — 60 million years ago

▲ Fossils give evidence for the evolution of the horse

C Describe how the fossils of dinosaur bones formed in rocks.

D How do scientists work out how organisms have changed over time?

How useful are fossil records?

It is rare for fossils to form, because the conditions must be right. Many fossils that have formed have not been found. Some fossils have been damaged, so they are less reliable. The evidence of past organisms in fossils is not complete.

Soft-bodied organisms and single-celled organisms, such as the earliest organisms, do not fossilise well. Any traces of such small organisms are often destroyed by geological processes such as erosion. This means that scientists are not absolutely certain about how early life forms evolved.

Questions

1 Explain how fossils help us to describe how an animal has evolved. ↓ E

2 (a) How do we know that dinosaurs existed? ↓ C

  (b) What evidence would scientists use to show that the dinosaurs were closely related to modern day reptiles?

3 Why do you think scientists can explain how the horse evolved, but have difficulty in reliably explaining how birds evolved? ↓ A*

Learning objectives

After studying this topic, you should be able to:

- ✔ understand extinction
- ✔ offer reasons for the causes of extinction
- ✔ understand what is meant by an endangered species

Key words

endangered, extinction

A What does the word 'extinction' mean?

B What is the difference between an endangered species and an extinct species?

Where are they now?

Evolution leads to new species being formed. These species evolve because they are better adapted to survive in their environment. Some existing species may not be able to compete or survive as well as the newer ones. The result is that they reduce in number, eventually becoming **endangered** as their numbers become low. If conditions do not improve for them, then all the members of the species will die out. This is called **extinction**. In this way, new species replace old ones in the cycle of life. One well known example of extinction is the dinosaurs.

The causes of extinction

There are many possible causes of extinction. Several factors may combine to bring about an extinction.

Changes to the environment over geological time

If the environment changes, for example the temperature increases, some organisms will not be as well adapted to the new conditions. They may be outcompeted by better-adapted rivals. The woolly mammoth became extinct about 10 000 years ago, partly because the climate became warmer.

▲ The woolly mammoth was widespread in northern regions of America and Europe, but became extinct about 10 000 years ago

Major catastrophic events

Major global events can affect the environment so much that some species cannot survive. Examples include massive volcanic eruptions and collisions with asteroids – one theory is that an asteroid impact resulted in reduced sunlight levels, which meant that plants could not photosynthesise so effectively. Many scientists believe that this led to the extinction of the dinosaurs.

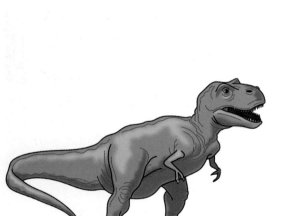

▲ Drawing of the extinct dinosaur *Tyrannosaurus rex*

New predators

The arrival of a new efficient predator might cause the extinction of some prey species. Humans are a good example – they can hunt as a team, and are very effective. Humans have been responsible for the extinction of many animals, such as the dodo.

◀ The dodo lived on the island of Mauritius. The birds became extinct in the seventeenth century, not long after humans first landed there. Humans hunted them, and rats released by humans ate their eggs.

▲ Dutch elm disease has wiped out more than 25 000 trees in the UK since 1967

New diseases

Some diseases are so virulent that they can wipe out an entire species before it has time to develop immunity. The elm tree population is almost extinct due to a fungal disease called Dutch elm disease.

New competitors

The arrival of a new and more effective competitor can cause problems for a resident species. An example is the introduction of the grey squirrel into the UK. This species outcompetes its relative the red squirrel, which is now in serious decline and is restricted to small areas of Scotland, Cumbria, Anglesey, and the Isle of Wight.

◀ The red squirrel has been largely replaced in the UK by the grey squirrel, which is more efficient at gaining resources such as food

Questions

1 What caused the dodo to become extinct?

2 Global warming is causing the icecaps to melt. What will be the effect on polar bear populations, and why?

3 Suggest why an asteroid impact might have led to the extinction of many plants at the time of the dinosaurs.

4 Explain how the introduction of the grey squirrel might lead to the eventual extinction of the red squirrel.

E
C
A*

Learning objectives

After studying this topic, you should be able to:

- ✔ know that during evolution new species form
- ✔ understand the mechanism by which a new species develops

Key words

speciation, isolation, **mutation**

▲ Alfred Russel Wallace

Did you know...?

Wallace was in a position to publish his ideas at about the same time as Darwin. If he had, a £10 note may now have a picture of Wallace instead of Darwin!

More and more new species

During evolution new species are formed. But how does this happen? Looking at the fossil record, it is clear to see that often one original ancestor species forms two new species. So the number of species increases.

Darwin knew that this must happen, but when he wrote *The Origin of Species* he could not explain how it happened. Modern biologists have explained the process.

Alfred Russel Wallace

Alfred Wallace was a Welsh biologist who worked at the same time as Darwin. He was also interested in explaining the evolution and formation of species. Although he did a lot of work collecting evidence, he was not as famous as Darwin. He found more evidence which contributed to Darwin's theory of evolution.

Alfred Wallace worked in South America and Malaysia. He was particularly interested in the ideas of geographical barriers leading to the formation of new species. He was perhaps the first biologist to suggest how isolating two populations might lead to **speciation**.

Forming new species

Sometimes two different populations of the same species change in different ways. Each group gradually changes over time, maybe to adapt to different environmental conditions. This then forms two new species. The process by which this happens is called speciation. It can be explained in four key steps.

> **A** What is speciation?

Step 1: isolation

The population becomes separated or **isolated** by some kind of barrier. The barrier could be a mountain range, a river, or the sea between different islands. Individuals in the two isolated populations of the species can no longer meet and interbreed, as they cannot cross the barrier.

Step 2: genetic variation

In each of the populations there will be variation between individuals. This variation is caused by a wide range of alleles. These are alternative versions of genes, leading to different characteristics. In addition, different **mutations** will occur in the separated populations. This increases the variation.

Step 3: natural selection

Over time, each separated group of the population evolves differently on each side of the barrier. Some of the variants will be better adapted to survive the conditions. Because conditions will be slightly different in the two areas, different variants will survive in each area. Each successful variant will pass on its own alleles to the next generation. The longer they are separated, the more different the two groups become.

Step 4: speciation

The two sub-populations have changed so much that they can no longer interbreed. They have formed two new closely related species.

▲ Modern humans and chimpanzees both evolved from a common ancestor over 5 million years ago in Africa. The ancestral population split into two groups, one standing upright and evolving into the modern human, the other becoming the modern chimp.

B What are the two main causes of variation in a species that, in the right conditions, could lead to speciation?

Questions

1 List three ways in which populations can become isolated. ↓E

2 Explain why two new species might not develop without a physical barrier.

3 Wallace discovered two species of monkey on either side of the river Amazon in South America. How could Darwin and Wallace have explained how they evolved from a common ancestor? ↓C

4 Over 65 million years ago, ancestral primates reached the island of Madagascar from mainland Africa.

  (a) Why did primates evolve into new primate species (even humans) in Africa, but into lemurs on Madagascar? ↓A*

  (b) What might have led to the evolution of more than 90 species of lemur on Madagascar?

Exam tip

✔ Learn the steps in the formation of a new species in sequence.

Course catch-up

Revision checklist

- Proteins make up much of the structure of living organisms and perform vital functions in cells.
- Enzymes are proteins that catalyse chemical reactions in cells. Each enzyme catalyses a specific reaction and works best at a particular temperature and pH.
- Enzymes are crucial for digestion.
- Biological washing powders have enzymes to break down stains.
- Enzymes are used in industry. They make baby food easier for babies to digest; can make sugar from cheap starch, for use in processed foods; some can turn glucose into a sweeter sugar called fructose.
- Respiration in cells provides the energy needed for an organism's life processes.
- Aerobic respiration requires oxygen.
- Anaerobic respiration does not require oxygen.
- Cells divide by mitosis for growth and asexual reproduction. Mitosis produces cells that are genetically identical to each other and to the parent cell. Living organisms made of many cells can be large, complex, and can have different types of cell for different functions.
- Certain cells divide by meiosis to make gametes for reproduction. Meiosis produces genetic variation.
- Gregor Mendel discovered how characteristics are inherited.
- Mendel's ideas explain many inheritance patterns. We use genetic diagrams to explain the results of genetic crosses.
- You are female if you inherit two X chromosomes, and male if you inherit one X and one Y chromosome.
- Genes are made of DNA and are found on chromosomes. Genes control characteristics because they contain coded information for making particular proteins.
- There are different versions of each gene. Some versions make abnormal proteins that lead to genetic disorders.
- Differences in people's DNA can be used to identify them. Embryos may be screened for genetic defects. Stem cells may be used to grow any type of cell to treat some illnesses.
- Fossils provide evidence for evolution and allow us to see how organisms have changed over time.
- Fossils show us what extinct organisms looked like.
- Mutations, isolation of populations, and competition all contribute to the formation of new species.

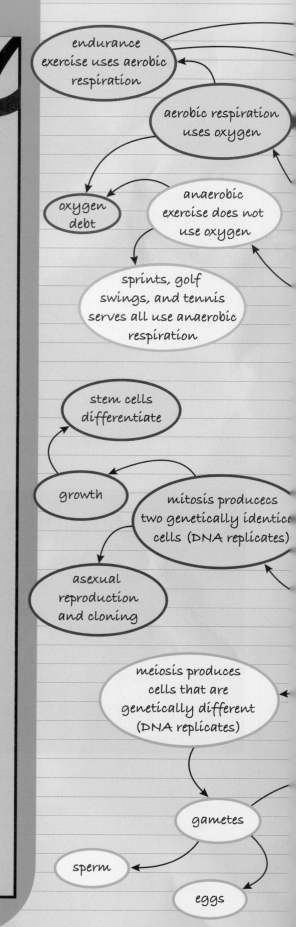

endurance exercise uses aerobic respiration

aerobic respiration uses oxygen

oxygen debt

anaerobic exercise does not use oxygen

sprints, golf swings, and tennis serves all use anaerobic respiration

stem cells differentiate

growth

mitosis producecs two genetically identica cells (DNA replicates)

asexual reproduction and cloning

meiosis produces cells that are genetically different (DNA replicates)

gametes

sperm

eggs

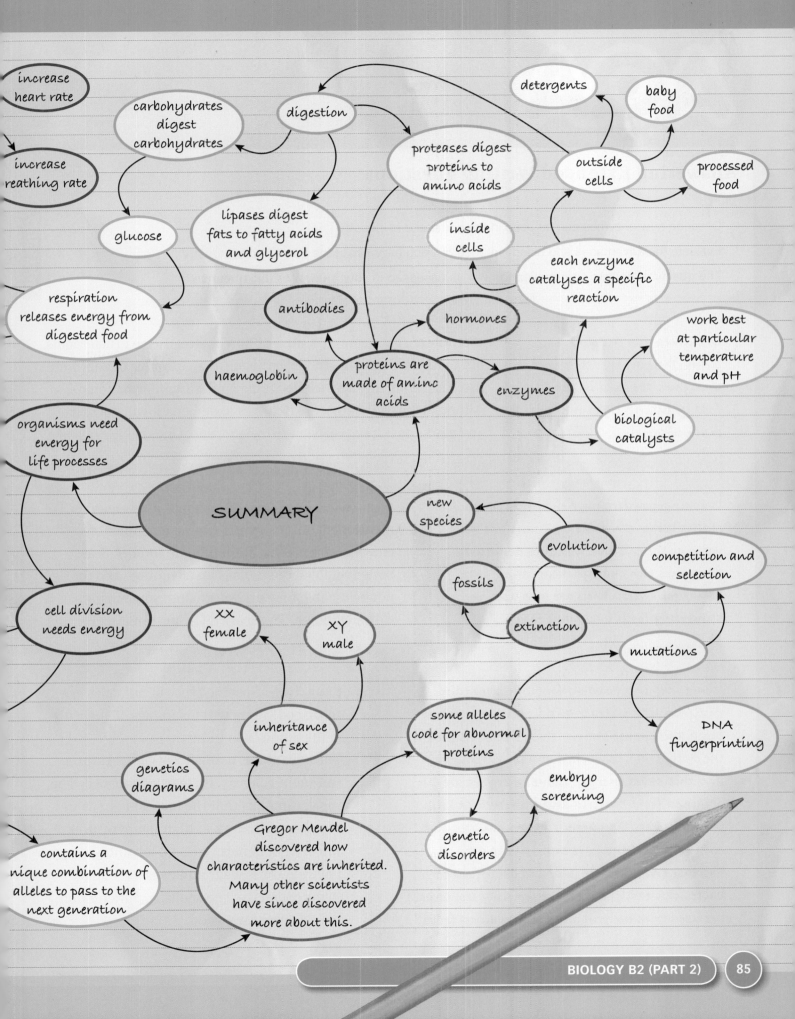

increase heart rate

increase reathing rate

carbohydrates digest carbohydrates

digestion

detergents

baby food

processed food

proteases digest proteins to amino acids

outside cells

glucose

lipases digest fats to fatty acids and glycerol

inside cells

each enzyme catalyses a specific reaction

respiration releases energy from digested food

antibodies

hormones

work best at particular temperature and pH

haemoglobin

proteins are made of amino acids

enzymes

organisms need energy for life processes

biological catalysts

SUMMARY

new species

evolution

competition and selection

fossils

cell division needs energy

XX female

XY male

extinction

mutations

inheritance of sex

some alleles code for abnormal proteins

DNA fingerprinting

genetics diagrams

embryo screening

contains a unique combination of alleles to pass to the next generation

Gregor Mendel discovered how characteristics are inherited. Many other scientists have since discovered more about this.

genetic disorders

Answering Extended Writing questions

Describe how enzymes can be used in the home and in industry.

Evaluate the advantages and disadvantages of using enzymes in the home and in industry.

The quality of written communication will be assessed in your answer to this question.

G–E

Enzymes help get cloths clean. They are in detergents. Enzymes are used in the food industry. Enzymes are made of protein.

Examiner: This answer is vague. There is a reference to detergents, but the term 'food industry' is too vague as an example of enzyme use. The candidate has included some irrelevant information. Although the last sentence is a correct statement, it will not score any marks as it is not relevant to the question. One spelling mistake.

D–C

Washing powders have enzymes to remove stains better. On some you cant use a hot wash as the enzyme would be denatured. Enzymes can change cheap starch into sugar for sweets. they are also used in slimming foods and baby food.

Examiner: Four uses of enzymes are briefly mentioned, but no details and no examples of enzymes are given. The candidate implies that better removal of stains is an advantage and that not being able to use a hot wash in some instances is a disadvantage. There are a few grammatical errors.

B–A*

Some washing powders for clothes contain enzymes like protease and lipase to get rid of blood and grease stains. These enzymes mean the clothes can be washed at a lower temperature and this saves energy. Some enzymes are obtained from bacteria and work well at high temperatures. They aren't denatured like most enzymes. But they are expensive to get. Protease enzymes are added to baby food so babies can easily digest the proteins. Isomerases are used to make fructose for slimming foods.

Examiner: Specific types of enzymes are described, along with examples of their use in the home or in industry.
One advantage, using lower temperatures and saving energy, is clearly described. The answer implies that the fact that enzymes from bacteria work at high temperatures and are not denatured is an advantage. This answer is clear and uses technical terms correctly. The spelling, punctuation, and grammar are good.

Exam-style questions

1 Scientists compared the performance of a new detergent with an existing detergent.

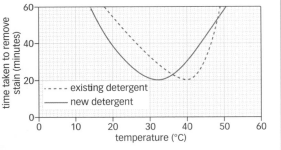

A03   **a**   Describe the effect of increased temperature on the time taken by the existing detergent to remove the stain.

A03   **b**   Which product works best at 30 °C?

A02   **c**   Explain why neither detergent works well at 60 °C.

A03   **d**   Is the new detergent likely to be more environmentally friendly than the existing detergent? Give a reason for your answer.

2   **a**   Why do living organisms need energy?
A01

A01   **b**   Write a word equation for aerobic respiration.

A01   **c**   Where in cells does most aerobic respiration take place?

A01   **d**   Write a word equation for anaerobic respiration.

A01   **e**   Which type of respiration releases more energy from each molecule of glucose?

A01   **f**   Which type of respiration causes muscle fatigue?

A02   **g**   Explain why breathing rate and heart rate increase when you run.

G–E

D–C

3 A student repeated one of Mendel's plant breeding experiments. She crossed pea plants grown from green seed with pea plants grown from yellow seed. All the first generation offspring produced yellow seeds. She crossed some of these with plants grown from green seed. The offspring from this second generation contained 19 plants with yellow seeds and 21 plants with green seeds.

A03   **a**   Which seed colour is the dominant characteristic?

A03   **b**   What is the ratio of yellow: green seeds in the second generation?

A02   **c**   Seed colour in peas is controlled by a single gene that has two alleles. Draw a genetic diagram to show why this ratio of yellow seeds to green seeds was produced in the second generation. Use A to represent the dominant allele and a to represent the recessive allele.

A02   **d**   How could the student have made the results more reliable?

B–A*

Extended Writing

4 Describe how enzymes in your digestive system digest food.
A01

G–E

5 Explain how fossils provide evidence for evolution.
A02

D–C

6 Describe how mitosis and meiosis are (a) similar   (b) different.
A02

B–A*

C2 Part 1

Structures, properties, and uses

Why study structures, properties, and uses?

Scientists have been uncovering the secrets of structures for centuries. We know what happens deep inside atoms. We know how particles give substances their properties. We know how to manipulate structures to make substances with perfect properties for particular purposes.

Today, scientists are excited about nanoscience – the study of structures measuring around one billionth of a metre. Will discoveries at the nanoscale lead to new cancer cures, greener energy technologies, and better computers?

In this unit you will learn how particles are arranged and joined together in metals, non-metals, and polymers, and about nanostructures. You will discover how the structures of substances influence their properties and uses. You will also learn about analysis, and how to calculate yields.

You should remember

1 Everything is made up of tiny particles, called atoms.

2 Protons, neutrons, and electrons make up atoms.

3 Atoms can share electrons to form covalent bonds.

4 Atoms can join together by losing or gaining electrons to form ions, which form ionic bonds.

5 The properties and uses of substances depend on how their particles are arranged.

6 Chromatography separates the components of a mixture, and can help us to identify the components.

7 Chemical formulae give information about the types and numbers of atoms that make up a substance.

8 Word equations show the reactants and products involved in chemical reactions.

The picture is an electron micrograph of the end of a nanotube. Nanotubes consist of sheets of carbon atoms arranged in hexagons. The sheets are wrapped around each other to form a cylinder with a hollow core. Ten thousand nanotubes side by side would only be as wide as a human hair.

Nanotubes are 10 000 times stronger than steel, but have a much smaller density. Added to tennis racquets, they already give elite players more power and control. Filled with metals, they make the world's smallest bar magnets. In future, nanotubes might make miniature circuits, or ultra-lightweight planes. The possibilities are endless...

1: Covalent bonding

Learning objectives

After studying this topic, you should be able to:

- ✔ explain how covalent bonds are formed
- ✔ draw dot and cross diagrams for simple molecules

Key words

compound, molecular formula, molecule, covalent bond, dot and cross diagram

Versatile gas

What do nitrogen fertiliser, explosives, household cleaner, and fish have in common?

These are all things that rely on ammonia. Ammonia is a raw material for making fertilisers and explosives. Ammonia solution is a household cleaner. Fish produce ammonia as a waste product. They excrete the ammonia into the water from their gills. Ammonia has a pungent smell – like toilets that need cleaning.

Inside ammonia

Ammonia is a **compound**. A compound is a substance that is made up of two or more different elements, chemically combined. So in a compound, the elements are not just mixed up. Chemical bonds join the elements together.

The **molecular formula** of ammonia is NH_3. This shows that ammonia is made up of atoms of two elements – nitrogen and hydrogen. There are three atoms of hydrogen for every one atom of nitrogen.

A The formula of water is H_2O. Name the two elements in this compound.

B Write down the formulae that represent the compounds in this list: H_2 Cl_2 CH_4 O_2 SiO_2

Joining atoms together

Ammonia gas exists as **molecules**. A molecule is a particle made up of two or more atoms chemically bonded together. In ammonia, each molecule consists of one atom of nitrogen joined to three atoms of hydrogen. The atoms are held together by **covalent bonds**. A covalent bond is a shared pair of electrons.

Covalent bonds form so that atoms can achieve stable electron arrangements. For example, a nitrogen atom has seven electrons. The electrons are arranged like this:

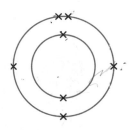

▲ Electronic structure of nitrogen   ▲ Electronic structure of hydrogen

There are five electrons in the highest occupied energy level, or outermost shell. This arrangement is not very stable. The nitrogen atom needs three more electrons in its highest energy level. It will then have the stable electronic structure of a noble gas, with eight electrons in its highest energy level.

A hydrogen atom has just one electron. It needs one more electron to achieve the stable electronic structure of the noble gas helium.

When nitrogen and hydrogen react to form ammonia, the two types of atom join together by *sharing* pairs of electrons. Each shared pair of electrons is one covalent bond. By sharing electrons, both the nitrogen and hydrogen atoms achieve stable electronic structures.

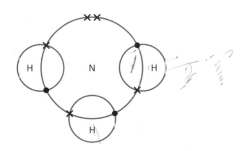

▲ In this **dot and cross diagram** for ammonia, crosses represent electrons from the nitrogen atom and dots represent electrons from the hydrogen atom. Each covalent bond is shown in a shared area as a dot and a cross.

Questions

1. How many atoms of each element are there in:
   (a) methane, CH_4?
   (b) ethanol, C_2H_5OH?
2. What is a molecule?
3. What is a covalent bond?
4. Draw dot and cross diagrams for:
   (a) water, H_2O
   (b) hydrogen chloride, HCl.

↓ E

▼ C

↓ A*

Exam tip

✔ When you draw dot and cross diagrams for covalent compounds, you only need to show the electrons in the highest occupied energy level (outermost shell).

2: More about molecules

Learning objectives

After studying this topic, you should be able to:

✔ describe covalent bonding in simple molecules

✔ draw displayed formulae for simple molecules

▲ Cows release huge quantities of methane to the atmosphere

Key words

displayed formula, double covalent bond, giant covalent structure, macromolecule

More molecules

Every day, rice crops release huge quantities of methane gas to the atmosphere. So do burping cows. Methane is a greenhouse gas. Its presence in the atmosphere contributes to global warming.

The central carbon atom in a methane molecule shares its outermost electrons with four hydrogen atoms. Each shared pair of electrons is a covalent bond. Here is a dot and cross diagram for methane:

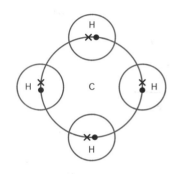

You can also represent the shared pairs of electrons like this:

In the diagram below, each line represents a shared pair of electrons. This is the **displayed formula** for methane. It shows the atoms in a molecule *and* the covalent bonds between them.

> **A** What is a displayed formula?
>
> **B** Draw displayed formulae for ammonia, NH_3, and water, H_2O.

Bonding in elements

There are covalent bonds in elements, too. The two atoms in a chlorine molecule, Cl_2, share a pair of electrons to make a single covalent bond.

The two oxygen atoms of an oxygen molecule share two pairs of electrons. This gives each oxygen atom the stable electronic structure of a noble gas, with eight electrons in the highest occupied energy level of each atom. The two shared pairs of electrons form a strong **double covalent bond**.

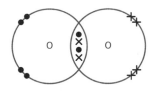

To find out about more about the properties of substances made from simple molecules, see spread C2.6.

Macromolecules

Some covalently bonded substances are joined together in huge networks called **giant covalent structures**, or **macromolecules**.

Diamond is a form of the element carbon. In diamond, covalent bonds join each carbon atom to four other carbon atoms. The bonds are arranged like this:

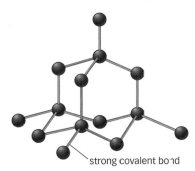

strong covalent bond

Silicon dioxide – the compound in most sand – also exists as macromolecules. Its atoms are joined together in a pattern like this:

key
- ○ silicon atom
- ● oxygen atom

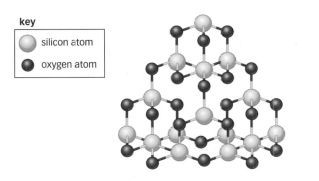

To find out about more about the properties of substances made from macromolecules, see spread C2.8.

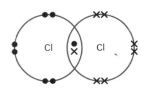

▲ This dot and cross diagram shows the electrons in the highest occupied energy levels of a chlorine molecule, Cl_2

▲ This dot and cross diagram shows the electrons in a hydrogen molecule, H_2

Exam tip **AQA**

✔ There are covalent bonds in compounds of non-metals and in non-metal elements.

Questions

1 What information is shown by a displayed formula?

2 Name two substances that have a giant covalent structure.

↓ E

3 Draw a displayed formula for a chlorine molecule.

↓ C

4 Explain why the two oxygen atoms in an oxygen molecule are joined by a double covalent bond.

5 How many covalent bonds does each silicon atom have in silicon dioxide?

↓ A*

Learning objectives

After studying this topic, you should be able to:

✔ explain what ions are

✔ explain how positive and negative ions are formed by electron transfer

✔ use dot and cross diagrams to represent ions

✔ describe ionic bonding

Vital ion

Kerry runs a marathon. At the end of the race, she vomits. She is dizzy and confused. In hospital, the doctor tells Kerry she drank too much water. The concentration of sodium **ions** in her blood is too low. She needs an injection of sodium chloride solution to replace the lost sodium ions – now!

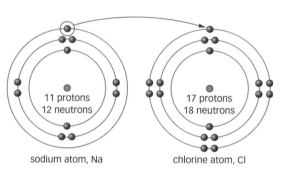

sodium atom, Na   chlorine atom, Cl

🔺 When sodium and chlorine react together, each sodium atom transfers one electron to a chlorine atom

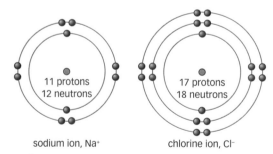

sodium ion, Na⁺   chlorine ion, Cl⁻

🔺 The diagrams show the electron arrangements in sodium and chloride ions

A Explain how positively charged ions are formed.

B Neon is a noble gas. How many electrons are in its highest occupied energy level?

Making ions

An ion is an electrically charged atom, or group of atoms. Ions form when atoms lose or gain electrons. Electrons are negatively charged, so:

- If an atom loses one or more electrons, it becomes a positively charged ion.
- If an atom gains one or more electrons, it becomes a negatively charged ion.

A sodium ion forms when a sodium atom transfers one of its electrons to an atom of a non-metal element, such as chlorine, in a chemical reaction:

- A sodium ion has 11 protons and 10 electrons. Overall, it has a charge of +1. Its formula is Na^+. You can also represent its electronic structure as $[2,8]^+$.
- A chloride ion has 17 protons and 18 electrons. Overall, it has a charge of –1. Its formula is Cl^-. You can also represent its electronic structure as $[2,8,8]^-$.

Each ion has eight electrons in its highest occupied energy level, like a noble gas. These electron arrangements are very stable.

Ionic bonding

Compounds made up of ions are called **ionic compounds**. There are strong electrostatic forces of attraction between the oppositely charged ions. This is **ionic bonding**. Ionic bonds act in all directions and hold the ions in a regular pattern, called a **giant ionic lattice.** See spread C2.7 to find out about the properties of ionic compounds.

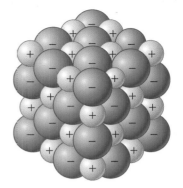

▲ The lattice structure of sodium chloride

▲ The aurora borealis

Did you know...?

Ions from the Sun stream towards the Earth and interact with its magnetic field, especially near the poles. Different substances in the upper atmosphere glow in different colours, giving a spectacular natural light show in the sky. This is known as the aurora borealis.

Key words

ion, ionic compound, ionic bonding, giant ionic lattice

Questions

1 What is an ion?
2 What is an ionic compound?
3 Describe the bonding in an ionic compound.
4 Draw and annotate a diagram to explain how a sodium atom transfers electrons to a chlorine atom to make ions.
5 Explain how ionic and covalent bonding are different.

Exam tip

- ✔ When a chlorine atom gains an electron, it forms a chloride ion, Cl^-.
- ✔ Covalent bonding involves sharing electrons, but ionic bonding involves transferring electrons.

Learning objectives

After studying this topic, you should be able to:

✔ describe the reactions of the Group 1 metals with non-metal elements

✔ represent the electronic structures of ions in ionic compounds

▲ In March 1930, Mahatma Gandhi led a non-violent protest against the British salt tax in colonial India

◀ Sodium reacts vigorously with chlorine to form sodium chloride

White gold

Thousands of years ago, humans discovered that salt preserves food. Salt became valuable. The Romans built roads just to transport it, and salt taxes made rulers rich. Wars have been fought over what we think of as an everyday item!

Making salt

Table salt is sodium chloride. It is made up of the elements sodium and chlorine. The salt we eat comes from the sea, or from underground rock salt. But you can make it in chemical reactions, too.

Flo's teacher heats a piece of sodium. He puts a gas jar of chlorine over it. There is a bright orange flame. Flo sees white clouds. Tiny white crystals of sodium chloride form on the sides of the gas jar.

Sodium is a metal. It is in Group 1 of the periodic table, the **alkali metals**. Each sodium atom has one electron in its highest occupied energy level, like all the other alkali metals. Chlorine is a non-metal. It is in Group 7 of the periodic table, the **halogens**. Each chlorine atom has seven electrons in its highest occupied energy level, like all the other halogens.

																		0
1	2						H					3	4	5	6	7		He
Li	Be											B	C	N	O	F	Ne
Na	Mg											Al	Si	P	S	Cl	Ar
K	Ca	Sc	Ti	V	Cr	Mn	Fe	Co	Ni	Cu	Zn	Ga	Ge	As	Se	Br	Kr
Rb	Sr	Y	Zr	Nb	Mo	Tc	Ru	Rh	Pd	Ag	Cd	In	Sn	Sb	Te	I	Xe
Cs	Ba	La	Hf	Ta	W	Re	Os	Ir	Pt	Au	Hg	Tl	Pb	Bi	Po	At	Rn
Fr	Ra	Ac														the halogens	

alkali metals

When sodium reacts with chlorine, each sodium atom transfers an electron to a chlorine atom. Two types of ions form – Na^+ and Cl^-. These ions make up the compound sodium chloride.

A Give the name and number of sodium's group in the periodic table.

B How many electrons are in the highest occupied energy level of a chloride ion?

Key words

alkali metal, halogen, metal halide, oxide ion

Alkali metals and the halogens

All Group 1 elements react vigorously with Group 7 elements, particularly if the metal is heated first. Clouds of a **metal halide** are produced. Fluorine produces a metal fluoride, chlorine produces a metal chloride, bromine produces a metal bromide, and iodine produces a metal iodide. The metal halide produced depends on the metal used, too. For example:

$$\text{lithium} + \text{bromine} \rightarrow \text{lithium bromide}$$
$$2Li + Br_2 \rightarrow 2LiBr$$

In all these reactions, each metal atom transfers one electron to a halogen atom. Two types of ion form:

- one with a single positive charge, such as Li^+
- one with a single negative charge, such as Br^-.

Alkali metals and oxygen

Alkali metals react with other non-metals, too. For example, heating sodium in air makes sodium oxide.

$$\text{sodium} + \text{oxygen} \rightarrow \text{sodium oxide}$$
$$4Na + O_2 \rightarrow 2Na_2O$$

Each oxygen atom has gained one electron from each of two sodium atoms. The sodium and oxide ions now have the stable electronic structures of noble gases. The formulae of the ions in sodium oxide are Na^+ and O^{2-}. It takes two sodium ions, Na^+, to balance the two charges on an **oxide ion**, O^{2-}, so the formula of sodium oxide is Na_2O.

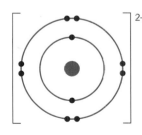

▲ You can represent the electronic structure of an oxide ion like this

Exam tip

✓ There is no overall charge on an ionic compound. The formula of an ionic compound shows that the number of positive charges is equal to the number of negative charges.

Questions

1 Give the name and number of fluorine's group in the periodic table.

2 How many electrons are in the highest occupied energy level of a sodium ion?

3 Name the product of the reaction of lithium with bromine.

4 Magnesium and calcium are in Group 2 of the periodic table. They have two electrons in their highest occupied energy level. Draw dot and cross diagrams for:

  (a) magnesium oxide, MgO

  (b) calcium chloride, $CaCl_2$.

E

C

A*

Learning objectives

After studying this topic, you should be able to:

✔ describe metallic bonding

▲ Lead is an excellent roofing material

A For a lump of lead, name the type of particle that is arranged in a regular pattern and the type of particle that is delocalised.

B Explain why electrons from the highest occupied energy level become delocalised in metals.

Exam tip **AQA**

✔ Be sure to mention delocalised electrons and positive ions when explaining metallic bonding.

Rogues on the roof

It's midnight. Police are called to a Bristol church. Thieves are taking lead from the roof. But why? And what was lead doing on the roof in the first place?

Lead is a metal. It is bendy and waterproof, and an excellent roofing material. The thieves wanted to sell the lead as scrap, for £650 a tonne. The scrap lead might end up in car batteries, protecting underwater cables, or even on another church roof.

Metallic bonding

Lead is a metal. Like all metals, it consists of a giant structure of atoms arranged in a regular pattern. Strong forces hold the structure together.

In metal atoms, the electrons in the highest occupied energy level are not strongly attracted to the nucleus. These electrons leave their atoms. They are **delocalised**, and free to move throughout the whole metal structure.

The atoms that have lost electrons are positive ions. They are arranged in a regular pattern to form a **giant metallic structure**. There are strong electrostatic forces of attraction between the positive ions and the moving delocalised electrons. This is **metallic bonding**.

To find out how metallic bonding influences the properties of metals, see spread C2.10.

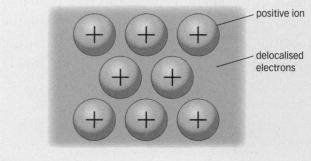

positive ion

delocalised electrons

▲ Metallic structure

How many electrons?

Potassium is in Group 1 of the periodic table. Its atoms have one electron in the highest occupied energy level. So potassium metal consists of a regular pattern of ions with a single positive charge, K^+. Delocalised electrons move throughout the metal. Attractive forces between the metal ions and the delocalised electrons hold the structure together.

Magnesium is in Group 2. Its atoms have two electrons in the highest occupied energy level. So magnesium metal contains Mg^{2+} ions arranged in a regular pattern. There are two delocalised electrons for every magnesium ion.

Key words

delocalised, giant metallic structure, metallic bonding

▲ A Roman lead water pipe at the Roman Baths in Bath

Did you know...?

The Romans smelted about 80 000 tonnes of lead each year. They used it to make water pipes and added it to silver coins to make them less valuable.

Questions

1 What is a delocalised electron?
2 Explain why metals contain positive ions.
3 Describe metallic bonding.
4 Use ideas about metallic bonding in sodium and magnesium to suggest why magnesium has a higher boiling point than sodium.

E
C
A*

6: Molecules and properties

Learning objectives

After studying this topic, you should be able to:

✔ explain how the properties of simple molecular substances are linked to their structure

 Bromine is a volatile liquid at room temperature

Exam tip AQA

✔ Remember – there are strong covalent bonds *within* molecules and weaker intermolecular forces *between* a molecule and its neighbours.

Key words

intermolecular forces

Bad smell

Guess which element is named from a Greek word meaning *to stink*?

The answer is bromine. Bromine is corrosive. Its vapours attack the eyes and lungs. Swallowing just 0.1 g of the liquid can be fatal. But bromine is not all bad. Its compounds are added to electronic goods and furniture to make them less likely to catch fire.

Inside bromine

Bromine consists of simple molecules. A simple molecule is made up of just a few atoms joined together by strong covalent bonds. The non-metal elements oxygen and hydrogen consist of simple molecules. So do the compounds hydrogen chloride, methane, and ammonia.

Properties of simple molecular substances

Substances that consist of simple molecules have low melting and boiling points compared to substances with ionic, metallic, or giant covalent structures.

Substance	Melting point (°C)	Boiling point (°C)
chlorine	−101	−35
zinc chloride	283	732
nitrogen	−210	−196
diamond	3550	4827
water	0	100
sulfur dioxide	−73	−10
zinc	420	907

A Look at the data in the table. Suggest which substances consist of simple molecules. Give reasons for your choices.

Explaining low melting and boiling points

Bromine is one of just two elements that are liquid at room temperature. It consists of bromine molecules, Br_2. Strong covalent bonds hold the atoms together in each bromine molecule.

The forces between a molecule and its neighbours are much weaker. It is these **intermolecular forces**, not the covalent bonds, that must be overcome when bromine melts or boils.

This explains why bromine has low melting and boiling points. Other substances that consist of simple molecules have low melting and boiling points too.

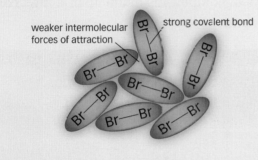

▲ The diagram shows the covalent bonds and intermolecular forces in liquid bromine

Substance	Boiling point (°C)
fluorine	−118
chlorine	−35
bromine	59
iodine	184

B Use the data in the table to describe the pattern in boiling points of the halogens.

C Explain why the halogens have low boiling points.

Did you know...?

Nitrogen gas becomes liquid at −196 °C. It is used to freeze blood and preserve genetic material, such as human eggs and sperm.

Good conductors?

Substances that consist of simple molecules do not conduct electricity. This is because the molecules do not have an overall electrical charge.

Questions

1 Describe three properties that are typical of substances that consist of simple molecules.

2 Explain why methane does not conduct electricity.

3 Predict which has the higher boiling point – copper or methane. Explain your prediction.

4 Use ideas about intermolecular forces to help you explain why ammonia has a low boiling point.

▲ Nitrogen gas condenses to become a liquid at −196 °C

Learning objectives

After studying this topic, you should be able to:

✔ explain how the properties of ionic compounds are linked to their structure

▲ The Taipei 101 building in Taiwan

Did you know...?

Magnesium oxide does not only protect building materials from fire. It relieves heartburn and stomach ache, and is an effective laxative.

Taipei 101

The Taipei 101 building in Taiwan is more than half a kilometre high. All the walls of its 101 storeys are covered with fire-resistant magnesium oxide. Magnesium oxide also protects the building's metal beams from fire.

Inside ionic compounds

Magnesium oxide is an ionic compound, like sodium chloride. It has a repeating structure of positive magnesium ions and negative oxide ions, arranged in a giant ionic lattice. There are strong electrostatic forces of attraction between the oppositely charged ions. The forces act in all directions.

Magnesium oxide and sodium chloride are not the only ionic compounds. Most compounds made up of a metal and a non-metal are ionic.

▲ Crystals of copper chloride (× 840)

Explaining the properties of ionic compounds

Ionic compounds have very high melting points. A scientist heats up solid magnesium oxide to 714 °C. The heat energy disrupts the strong forces of attraction in the giant ionic lattice. The regular pattern of ions breaks down. The magnesium oxide melts and becomes liquid.

The scientist heats the liquid magnesium oxide even more, to 1422 °C. At this temperature, there is enough energy to overcome the strong forces of attraction between the ions. The magnesium oxide liquid boils and becomes a gas.

All ionic compounds have high melting points and high boiling points because of the large amounts of energy needed to break their many strong bonds.

Electricity

Ionic compounds do not conduct electricity when they are solid. Their ions are not free to move from place to place to carry the current.

Ionic compounds conduct electricity when they are liquids, but their high melting points make this difficult to show in the laboratory. In ionic liquids, the ions are free to move from place to place to carry the current.

Ionic compounds also conduct electricity when they are dissolved in water. Again, the ions are free to move from place to place. Of course, not all ionic compounds are soluble in water.

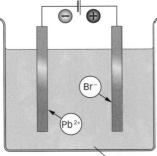

melted lead bromide

▲ Lead bromide is an ionic compound. Liquid lead bromide conducts electricity because its ions are free to move towards the electrodes.

A Describe the forces that hold the ions together in ionic compounds.

B Explain why magnesium oxide has a high boiling point.

Exam tip

✔ Solid ionic compounds do not conduct electricity. Liquid ionic compounds, and those dissolved in solution, do conduct electricity.

Questions

1 Under what conditions does sodium chloride conduct electricity?

2 Name the type of force that holds the ions together in an ionic compound.

3 Suggest why magnesium oxide bricks are used to line the inside of furnaces.

4 Explain why ionic compounds have high melting points.

5 Look at the boiling points in the table. Suggest which compound is not an ionic compound. Give a reason for your decision.

Name of compound	Boiling point (°C)
magnesium chloride	1412
lead bromide	916
sulfur dioxide	−10
rubidium iodide	1300
silver chloride	1550

Learning objectives

After studying this topic, you should be able to:

✔ recognise the structures of diamond and graphite, and describe their properties

✔ explain some of the uses of diamond and graphite

✔ explain the properties of diamond and graphite in terms of their structure

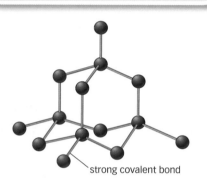

strong covalent bond

▲ Diamond structure

▲ Diamond's hardness and high melting point means it makes excellent dental drill bits

Crazy carbon

Diamond and **graphite** are different forms of the same element – carbon. Diamond is lustrous, hard, and makes stunning jewellery. Graphite is soft and grey. Mixed with clay, it makes pencil 'leads'. How can these two forms of the same element be so different? The answer is in their structures.

Diamond – the inside story

In diamond, each carbon atom forms strong covalent bonds with four other carbon atoms. The pattern is repeated to make a giant covalent structure.

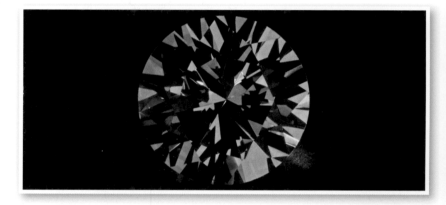

▲ There are about 10 000 000 000 000 000 000 000 carbon atoms in this 0.2 g diamond

The strong covalent bonds explain two important properties of diamond:

- its extreme hardness
- its high melting point of 3550 °C.

Diamonds are transparent. They can be cut into shapes that allow light to pass through them so that they seem to sparkle. Their lustre and hardness mean that diamonds are valuable gemstone jewels.

Diamond does not conduct electricity because it has no charged particles that are free to move.

> **A** Explain why diamond is hard.
>
> **B** Give one use of diamond that depends on its hardness. What other property of diamond makes it suitable for this purpose?

Graphite

The carbon atoms in graphite are arranged differently from those in diamond. Graphite has a layered structure. Each carbon atom is joined to three others by strong covalent bonds. But the forces *between* the layers are weak, so the layers can slide over each other easily. This is why graphite is slippery.

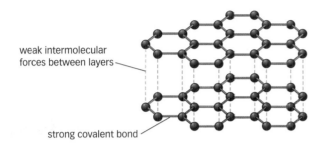

weak intermolecular forces between layers

strong covalent bond

▲ Graphite structure

Pencil 'lead' is not lead at all, but a mixture of graphite and clay. The black slippery graphite wears away on paper, leaving a black line.

▲ Pencil 'leads' are mainly graphite

Graphite's slipperiness also makes it useful in **lubricants**. At low temperatures, oil helps moving machine parts slide over each other easily. But oil breaks down at high temperatures. Graphite has a high melting point, and so makes a better high temperature lubricant than oil.

In graphite, one electron from each carbon atom is free to move. These delocalised electrons explain why graphite can conduct electricity. This property means that graphite is useful as **electrodes** in electrolysis. The delocalised electrons also allow graphite to conduct heat. The ability to conduct heat and electricity makes graphite like metals.

Key words

diamond, graphite, lubricant, electrode

Exam tip

✔ Be prepared to describe the different properties of diamond and graphite, and to link these properties to their structures and their uses.

Questions

1 Name the element of which graphite and diamond are different types.

2 Make a table to compare the properties of diamond and graphite.

↓ E

3 Explain why diamond is used in drills, cutting tools, and jewellery.

4 Explain why graphite is used in pencils, lubricants, and electrodes.

↓ C

5 In terms of their structure, explain why:

(a) Graphite conducts electricity but diamond does not.

(b) Graphite is slippery but diamond is hard.

↓ A*

(c) Graphite and diamond both have high melting points.

9: Bucky balls and nanotubes

Learning objectives

After studying this topic, you should be able to:

- ✔ explain what fullerenes are, and why they are useful
- ✔ explain the meanings of the terms nanoscience and nanoparticles
- ✔ describe some applications of nanoscience

Key words

fullerene, **buckminsterfullerene**, **nanoparticle**, **nanoscience**, **nanotube**

Did you know...?

Three scientists – Robert Curl, Harold Kroto, and Richard Smalley, discovered buckminsterfullerene in 1985. It was named after an American architect who designed 'geodesic domes'. These have the same sort of arrangement of hexagons and pentagons as buckminsterfullerene.

▲ The Eden project in Cornwall has giant geodesic domes

Fabulous fullerenes

Diamond and graphite are not the only forms of carbon. The element can also exist as **fullerenes**. Fullerenes are a type of carbon based on hexagonal rings of carbon atoms. Their properties are amazing!

The most common fullerene is **buckminsterfullerene**. It consists of molecules containing 60 carbon atoms joined together to form a hollow sphere. Its chemical formula is C_{60}.

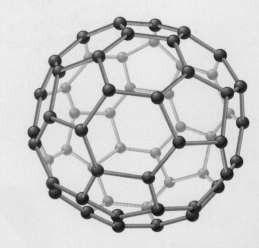

Fullerenes are hollow. The space inside is big enough for atoms and small molecules to fit in. Scientists have discovered how to 'cage' radioactive metal atoms and drug molecules inside fullerenes. These fullerenes can be used to deliver drugs into the body. For example, they can be coated with chemicals that make them gather next to cancer cells after being injected into the body. In this way, cancer drugs can be delivered to their targets without damaging normal cells.

A What are fullerenes?

B Explain why fullerenes can be used to deliver drugs into the body.

Nanoscience

Fullerenes are examples of **nanoparticles**. Nanoparticles are tiny particles made up of a few hundred atoms. They measure between about 1 nanometre and 100 nanometres across. A nanometre is one billionth of a metre, so nanoparticles are much too small to see with a microscope. **Nanoscience** is the study of nanoparticles.

Nanoparticles are proving to be incredibly useful, and scientists keep on thinking of new ways of using their unique properties. For example:

- Fullerene particles join together to make **nanotubes**. These have a huge surface area compared to their volume. So they make excellent catalysts, to speed up reactions.
- Nanotubes are the strongest and stiffest materials ever discovered. So they are useful for reinforcing graphite tennis racquets.

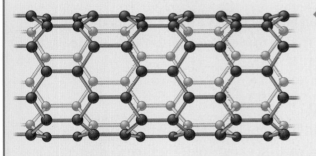

◀ The structure of a nanotube

The properties of nanoparticles are different from those of the same substances in normal-sized pieces. For example, titanium dioxide is a white solid. It is used in house paint. But titanium dioxide nanoparticles are so small they cannot reflect visible light. You cannot see them. This means they make excellent sun block creams. They protect your skin from harmful ultraviolet light without making your skin look white.

The amazing properties of nanoparticles could lead to many exciting developments in technology, including the development of

- new types of computer
- new coatings (in self-cleaning ovens and windows, for example)
- sensors that detect substances in tiny amounts
- very strong, light building materials.

Exam tip

✔ You need to know what is meant by nanoscience and nanoparticles, but you do not need to remember examples of particular nanoparticles or properties.

▲ This water droplet is resting on a water-repellent surface. The surface contains nanoparticles that increase the contact angle between it and the water. These surfaces can withstand corrosion and stay clean.

Questions

1 Explain the meanings of the words nanoparticle and nanoscience.

2 Describe three uses of nanoparticles.

3 Explain why nanotubes make good catalysts.

4 Explain why nanoparticles are added to the materials used to make some tennis racquets.

5 Describe some social and economic benefits of nanoscience.

Learning objectives

After studying this topic, you should be able to:

✔ explain how the properties of metals are linked to their structure

✔ explain why alloys have different physical properties to the elements from which they are made

✔ describe the properties and uses of shape memory alloys

Clever copper

Electricians choose copper for their cables. Copper is bendy, and a good conductor of electricity.

◀ Copper cables

Metallic structure explains the properties of copper and other metals. There is a diagram showing metallic structure on spread C2.5. The positively charged metal ions are packed tightly in layers. The layers slide over each other easily. This explains why metals can be bent and shaped.

> Metals conduct electricity because their delocalised electrons are free to move throughout the metal to carry the current.

Bling bling

Tanya buys gold fingernails to stick over her own nails. They are plated with 18 carat gold. Eighteen carat gold is not pure gold. Pure gold is too soft for false fingernails. It scratches easily and wears away quickly when it rubs against harder materials.

Eighteen carat gold is an **alloy**. An alloy is a mixture of a metal with one or more other elements. The physical properties of an alloy are different from the properties of the elements in it. Eighteen carat gold is a mixture of 75% gold mixed with copper and silver. The copper and silver atoms distort the layers in the gold structure, making it more difficult for them to slide over each other. This makes gold alloys harder than pure gold.

▲ Tanya has gold nails!

A Tanya sees an advert for getting her mobile phone plated with 24 carat gold, which is 99.9% gold. Suggest why 18 carat gold might be more suitable for this purpose.

B Use ideas about structure to explain why gilding alloy (95% copper and 5% zinc) is harder than pure copper.

◀ An alloy is a mixture of a metal with small amounts of one or more other elements

Smart alloys

Do you wear dental braces? If so, you have experienced **shape memory alloys** in action. Shape memory alloys – also called smart alloys – remember their original shapes. If a smart alloy is bent or twisted it keeps its new shape. But heating a smart alloy above a certain temperature makes the alloy return to its original shape.

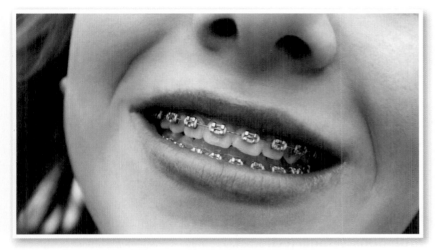

▲ Dental braces

Some of the best smart alloys are mixtures of nickel and titanium, called nitinol. Dental braces are made of nitinol wire. The wire exerts a constant force on the teeth to return the teeth to their correct positions.

Smart alloys have one main problem – they break easily if they are bent or twisted too much.

Key words

alloy, shape memory alloy

Did you know...?

Some glasses frames are made of nitinol. If you sit on them, you can gently warm the frame to return them to their original shape.

Questions

1 Name the bonds found in metals.
2 Explain why electrical wires are made from copper.
3 Describe two uses of smart alloys. Explain how their properties make them suitable for these purposes.
4 Explain why alloys are often harder than the metals that are mixed to make them.
5 Describe metallic bonding, and explain why metals are easy to bend and how they conduct electricity.

Exam tip

✔ In the exam, you need to be able to explain why a metal is suitable for a particular use. Sometimes, but not always, you may be given data to analyse.

Learning objectives

After studying this topic, you should be able to:

✔ explain how the properties of polymers are linked to what they are made from and the conditions under which they are made

✔ explain how the uses of polymers are linked to their structures

Two sorts of poly(ethene)

The bag and the bottle below are both made from poly(ethene). So why are their properties so different?

There are two types of poly(ethene). The bag is made from **low density poly(ethene)**, LDPE. The bottle is made from **high density poly(ethene)**, HDPE.

Each type of poly(ethene) has its own properties.

	LDPE	HDPE
density (g/cm³)	0.92	0.95
strength (MPa)	12	31
transparency	good transparency	less transparent
relative flexibility	flexible	stiff

The two types of poly(ethene) are both made from the same starting monomer – ethene. But they are made under different conditions.

	LDPE	HDPE
temperature (°C)	100–300	300
pressure (atm)	1500–3000	1
catalyst or initiator	oxygen or peroxide initiator	aluminium-based metal oxide catalyst

The properties of all polymers depend on what they are made from and the conditions under which they are made.

A Name two types of poly(ethene).

B Use the data in the first table to explain why it is better to make bottles from HDPE than LDPE.

Thermoplastic and thermosetting polymers

Some plastics, such as poly(ethene), soften easily when they are warmed. It is easy to mould them into new shapes, so they can be recycled. These are **thermosoftening polymers**. They consist of individual, tangled polymer chains.

> The forces of attraction between the separate polymer chains are weak.

weak forces between the separate polymer chains

▲ Polymer chains in a thermosoftening polymer

Some plastics cannot be recycled because they do not melt when they are heated. These are **thermosetting polymers**. Thermosetting polymers consists of polymer chains with cross-links between.

> These cross-links are strong intermolecular bonds.

chains held together by strong bonds

▲ Polymer chains in a thermosetting polymer

Key words

low density poly(ethene),
high density poly(ethene),
thermosoftening polymer,
thermosetting polymer

Questions

1 Describe three differences in the properties of LDPE and HDPE.

2 What is a thermosetting polymer?

3 Suggest some uses for LDPE and HDPE.

4 Explain why it is possible to melt thermosoftening polymers, but not thermosetting ones.

5 Draw a table to show the differences between thermosetting and thermosoftening polymers, and the reasons for these differences.

↓ E

▼ C

↓ A*

Exam tip

✔ Thermosoftening plastics melt because they consist of individual chains. Thermosetting plastics have cross-links, so cannot melt.

Learning objectives

After studying this topic, you should be able to:

✔ work out the mass number and atomic number of an atom

✔ explain what an isotope is

Key words

sub-atomic particle, nucleus, proton, neutron, electron, atomic number, mass number, isotope

Brain scan

Scans like this have unlocked some of the brain's deepest secrets. It is thanks to an understanding of atoms – and what goes on inside them – that scientists have been able to develop techniques like this.

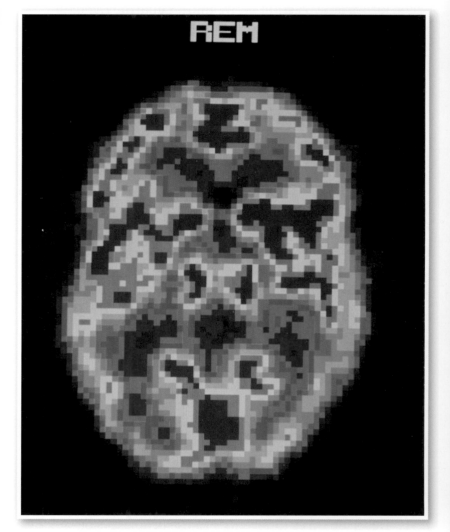

▲ A scan of a human brain

Inside atoms

Atoms are the smallest part of an element that can exist. The diameter of a typical atom is 0.00000001 cm. But it is made up of even tinier particles, called **sub-atomic particles**.

In the centre of an atom is the **nucleus**. The nucleus is made up of **protons** and **neutrons**. The nucleus is surrounded by **electrons**. The table shows the masses and charges of these sub-atomic particles.

Name of particle	Relative mass	Relative charge
proton	1	+1
neutron	1	0
electron	very small	−1

Identifying atoms

If you look for sodium on the periodic table, you will see it is represented like this:

$$^{23}_{11}\text{Na}$$

The number 11 is the **atomic number** of sodium. It is equal to the number of protons in a sodium atom, and also the number of electrons. The number 23 is the **mass number**. It is equal to the total number of protons and neutrons in a sodium atom. The number of neutrons is the mass number minus the atomic number. For example $^{23}_{11}\text{Na}$ contains 12 neutrons (23 – 11).

> **A** Work out the number of protons, neutrons, and electrons in an atom of $^{12}_{6}\text{C}$.
>
> **B** Work out the number of protons, neutrons, and electrons in an atom of $^{197}_{79}\text{Au}$.

Isotopes

Atoms of the same element can have different numbers of neutrons. This means they have different mass numbers. Atoms of the same element which have different numbers of neutrons are called **isotopes**.

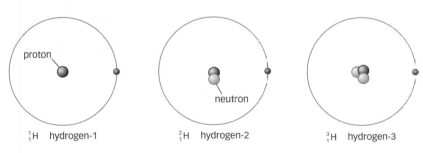

$^{1}_{1}\text{H}$ hydrogen-1 $^{2}_{1}\text{H}$ hydrogen-2 $^{3}_{1}\text{H}$ hydrogen-3

▲ Three different isotopes of hydrogen. They contain the same number of protons and electrons, but different numbers of neutrons.

Every chlorine atom contains 17 protons and 17 electrons. About 75% of chlorine atoms have 18 neutrons. The other 25% of chlorine atoms have 20 neutrons. So there are two isotopes of chlorine. One has a mass number of 35 (17 + 18) and the other has a mass number of 37 (17 + 20).

Exam tip **AQA**

✔ You will be given a copy of the periodic table in the exam.

Questions

1 What is the charge on a proton?

2 Give the relative mass of a proton and the relative mass of an electron.

3 Define the term 'atomic number'.

4 What is an isotope?

5 Write a table to show the numbers of protons, neutrons, and electrons in the seven naturally occurring isotopes of mercury: $^{202}_{80}\text{Hg}$ $^{200}_{80}\text{Hg}$ $^{199}_{80}\text{Hg}$ $^{201}_{80}\text{Hg}$ $^{198}_{80}\text{Hg}$ $^{204}_{80}\text{Hg}$ $^{196}_{80}\text{Hg}$.

↓ E

↓ C

↓ A*

13: Masses and moles

Learning objectives

After studying this topic, you should be able to:

✔ calculate the relative formula mass of a given substance

Did you know...?

A gold atom has a mass of 3×10^{-21} kg. That's just 0.000 000 000 000 000 000 003 kg.

▲ Atoms of gold piled on top of a layer of carbon atoms

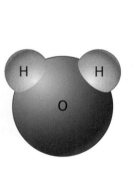

▲ The M_r of water is 18. Each water molecule has 1.5 times the mass of a carbon atom.

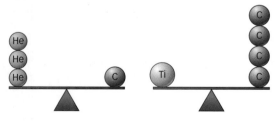

▲ Three helium atoms have the same mass as one carbon atom, and one titanium atom has the same mass as four carbon atoms

Relative atomic mass

Individual atoms are tiny. They have very little mass. The gold atoms in the photograph are less than a millionth of a millimetre in diameter. The atoms have such small masses that it is more useful to use their **relative atomic mass** rather than their actual mass in kilograms.

Carbon atoms are the standard atom against which all the others are compared. The relative atomic mass, A_r, of the most common carbon isotope is exactly 12. Atoms with an A_r of less than 12 have less mass than a carbon atom, and those with an A_r greater than 12 have more mass than a carbon atom.

The relative atomic mass of an element is an average value for the isotopes of the element, taking into account their relative amounts. For example, about 75% of chlorine atoms have a mass number of 35. The other 25% have a mass number of 37. The relative atomic mass of chlorine is 35.5, an average of the masses of the two isotopes, taking into account their relative amounts.

> **A** How many helium atoms have a total mass equal to the mass of one titanium atom?

Relative formula mass and moles

The chemical **formula** of a substance tells you the number of each type of atom in a unit of that substance. For example, the formula for water is H_2O. It shows that each water molecule consists of two hydrogen atoms and one oxygen atom, joined together.

The **relative formula mass** or M_r of a substance is the mass of a unit of that substance compared to the mass of one carbon atom. It is worked out by adding together all the A_r values for the atoms in the formula.

Scientists say that the relative formula mass of a substance, in grams, is one **mole** of that substance. So the mass of one mole of carbon atoms is 12 g.

Key words

relative atomic mass, formula, relative formula mass, mole,

Worked example 1

What is the M_r of water, H_2O?

A_r values: H = 1, O = 16

M_r of H_2O = 1 + 1 + 16 = 18

Worked example 2

What is the M_r of magnesium hydroxide, $Mg(OH)_2$?

A_r values: Mg = 24, O = 16, H = 1

M_r of $Mg(OH)_2$ = 24 + [2 × (16 + 1)]

$\qquad$ = 24 + 34 = 58

(Notice that the 2 outside the brackets in $Mg(OH)_2$ means that there are two oxygen atoms and two hydrogen atoms.)

Worked example 3

What is the mass of one mole of magnesium hydroxide?

M_r of $Mg(OH)_2$ = 58, so the mass of one mole = 58 g

B What is the relative formula mass of magnesium oxide, MgO?

Questions

Use the periodic table to find the answers to Questions 1 and 2.

1 What are the relative atomic masses of nitrogen, chlorine, and sodium?

2 What is the relative formula mass of:

$\qquad$ (a) oxygen, O_2

$\qquad$ (b) carbon dioxide, CO_2

$\qquad$ (c) ammonia, NH_3

$\qquad$ (d) sodium chloride, NaCl.

3 What is the relative formula mass of aluminium hydroxide, $Al(OH)_3$?

4 What is the relative formula mass of ammonium sulfate, $(NH_4)_2SO_4$?

↓ E

↓ C

↓ A*

Learning objectives

After studying this topic, you should be able to:

- ✔ describe how to use paper chromatography to identify food additives
- ✔ explain how gas chromatography separates the substances in a mixture

Identifying food colourings

Tamara buys some sweets. She wants to know what colourings they have in them. She sets up a **paper chromatography** investigation to separate the colourings. She obtains this **chromatogram**:

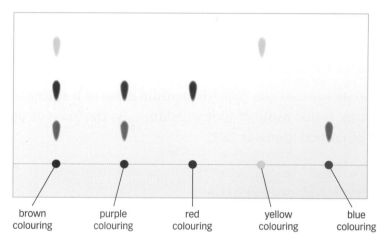

▲ The chromatogram shows that the brown colouring is a mixture of blue, red, and yellow dyes

In chromatography, a **mobile phase** moves through a **stationary phase**. Tamara added the sweet colouring sample to the stationary phase (paper). The mobile phase (in this case water) carried the chemicals in the sample through the stationary phase. Each compound moved at a different speed. So, the compounds in the mixture separated.

> **A** Which two food colourings are mixed in the purple dye?

Drunk driver?

A police officer stopped Darren, who was driving his lorry erratically. She thought Darren might be drunk. Darren went to the police station but the breath test machine was broken. So he gave a urine sample instead.

The police officer sent the urine sample to a forensic laboratory. At the laboratory, scientists used **gas chromatography** to measure the exact concentration of alcohol in the urine.

Gas chromatography

Gas chromatography is an instrumental method for detecting, identifying, and measuring chemical compounds.

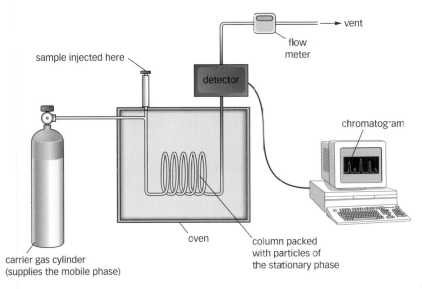

▲ Inside a gas chromatography instrument

Here's how it works:

- The urine sample is heated so that it becomes a mixture of vapours.
- A carrier gas, usually helium, mixes with the vapours. The gas is the mobile phase.
- The carrier gas takes the mixture of different vapours from the urine sample through a column packed with a solid material. The solid material in the column is the stationary phase.
- Different substances in the vapour mixture travel through the column, towards the detector, at different speeds. They become separated.

But was Darren over the limit? See spread 2.15 to find out how data from the gas chromatography analysis will provide the answer.

See spread 2.15

Key words

paper chromatography, chromatogram, mobile phase, stationary phase, gas chromatography

Did you know...?

The word 'chromatography' comes from the Greek word for colour.

Questions

1 What is the stationary phase in paper chromatography?

2 In paper chromatography, what happens to the substances in the mixture?

3 What is the purpose of gas chromatography?

4 Name the commonly used carrier gas in gas chromatography.

5 Describe the stages by which the compounds in a sample are separated in gas chromatography.

E

C

A*

Learning objectives

After studying this topic, you should be able to:

✔ explain how a combination of gas chromatography and mass spectrometry identifies the compounds in a mixture

Key words

retention time, mass spectrometer, molecular ion peak

▲ Darren went to court and lost his licence and his job

Interpreting gas chromatograms

Was Darren over the limit? Read on to discover the answer …

The gas chromatography column separates the substances in Darren's urine as they move towards the detector at the end of the column:

- The substance that travels most quickly reaches the detector first. This substance has the shortest **retention time**.
- The detector sends a signal to a recorder, which draws a peak on a chromatogram.
- The other substances reach the detector, one by one. The one that arrives last has the longest retention time.
- Scientists look at the chromatogram. Each peak represents one of the substances in the original mixture.
- The number of peaks shows the number of compounds present in the original mixture.
- The time taken for a substance to travel through the column helps to identify the substance.
- The relative areas under the peaks show the relative amounts of each of the compounds in the mixture.

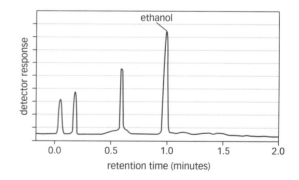

▲ A simplified version of part of the chromatogram from Darren's urine. It shows that there is ethanol (alcohol) in his urine. The scientists interpreted the data and said he was over the legal limit.

A A gas chromatogram has three peaks. What does this tell us about the mixture being analysed?

B Why is it useful to know the time taken for a substance to travel through a gas chromatography column?

Mass spectrometry

Often, the detector in gas chromatography is a **mass spectrometer**. The mass spectrometer identifies the separated substances from gas chromatography very quickly and accurately, and can detect very small quantities.

The mass spectrometer can also give the relative molecular mass of each of the substances separated in the column.

Mass spectrograph

Mass spectrometers produce mass spectrographs which look something like this.

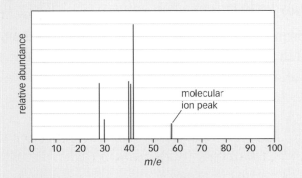

The peak on the right is the **molecular ion peak**. The mass of the molecular ion represents the relative formula mass of the molecule.

Advantages of instrumental methods

Like all instrumental methods for detecting and identifying elements and compounds, gas chromatography and mass spectrometry are

- accurate
- sensitive
- quick.

▲ Researcher analysing results from a mass spectrometer

Did you know...?

In 2010, the drink-drive limit in the UK was 107 mg of alcohol per 100 ml of urine. This is higher than that of most other European countries.

Exam tip

✔ Gas chromatography separates substances. Mass spectrometry identifies them.

Questions

1 On a gas chromatogram, what does the number of peaks show?

2 Which substance travels more quickly through a gas chromatography column – one with a shorter or one with a longer retention time?

3 Explain how a gas chromatogram indicates the relative amounts of the substances in a mixture.

4 Give three benefits of using instrumental methods to analyse mixtures.

5 Explain the purpose of using a mass spectrometer as the detector in gas chromatography.

↓ E

↓ C

↓ A*

Learning objectives

After studying this topic, you should be able to:

- ✔ calculate the percentage of an element in a compound
- ✔ calculate empirical formulae
- ✔ calculate masses from equations

A Calculate the percentage by mass of potassium in potassium hydroxide, KOH.

B Calculate the percentage by mass of nitrogen in ammonium nitrate, NH_4NO_3

▲ Many farmers add fertilisers to crops to improve their yields

Calculating the percentage by mass of an element in a compound

Plants need potassium ions. So potassium chloride fertilisers improve crop yields. You can calculate the **percentage by mass** of potassium in the compound like this:

- Write down the formula of the compound.   KCl
- Use A_r values to calculate the relative formula mass. $39 + 35.5 = 74.5$
- Divide the A_r of potassium by the M_r of KCl. $39 \div 74.5 = 0.52$
- Multiply by 100 to get a percentage. $0.52 \times 100 = \mathbf{52\%}$

If there is more than one atom of an element in the formula, you need to multiply the A_r value of that element by the number of atoms shown.

Calculating empirical formulae

If you know the masses of the elements in a sample of a compound, you can calculate the formula of the compound.

Worked example 1

A sample of a compound contains 1.2 g of carbon and 3.2 g of oxygen. What is its formula?

	carbon	oxygen
Mass of each compound	1.2 g	3.2 g
A_r (from periodic table)	12	16
mass divided by A_r	0.1	0.2
simplest ratio	1	2

So the formula of the compound is CO_2.

Calculating masses of reactants and products from equations

In a chemical reaction, atoms are not created or destroyed. So the total mass of reactants is equal to the total mass of products.

Key words

percentage by mass

Worked example 2

Catherine heats 200 g of calcium carbonate. It decomposes to make calcium oxide and carbon dioxide. How much calcium oxide does she make?

$$CaCO_3 \rightarrow CaO + CO_2$$

M_r of $CaCO_3 = 40 + 12 + (16 \times 3) = 100$

$= 40 + 16 = 56$

Use ratios to work out the answer:

100 g of calcium carbonate would make 56 g of calcium oxide, so

200 g of calcium carbonate makes 112 g of calcium oxide.

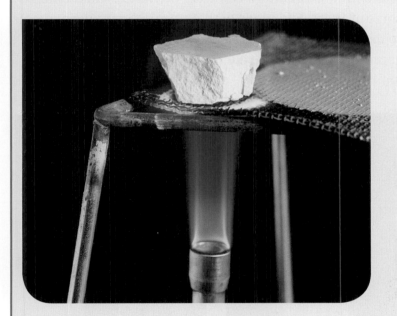

▲ Heating calcium carbonate

Questions

1 Calculate the percentage by mass of carbon in carbon dioxide, CO_2.

2 A sample of a compound contains 3.9 g of potassium, 1.2 g of carbon, and 1.4 g of nitrogen. What is its formula?

3 A compound is made up of 27% sodium by mass, 16% nitrogen by mass, and 56% oxygen by mass. What is its formula?

4 Edward heats 24 g of carbon in air. What mass of carbon dioxide does he make?

Learning objectives

After studying this topic, you should be able to:

✓ explain why it is not always possible to obtain the calculated amount of product

✓ calculate the percentage yields of reactions

✓ explain what a reversible reaction is

▲ Some product may be lost during filtration

A Give three possible reasons for the actual yield in a reaction being less than the maximum theoretical yield.

B Suggest an economic reason for chemists in a chemical company wishing to make as much product as possible.

Less than 100%

In most chemical reactions, chemists want to make as much product as possible. You can use reaction equations to calculate the maximum mass of a product you could expect to make (see spread C2.16). But, even though atoms are not created or destroyed in chemical reactions, you are likely to make a smaller mass than the calculated mass. This might be because:

*   Some of the product was lost when you separated it from the reaction mixture. For example, if you are separating a solid product from a solution, some of the solid product might get stuck in the filter paper.
*   The reactants may react in ways that are different from the expected reaction. For example, if you burn lithium in air to make lithium oxide, you might also make lithium nitride.
*   The reaction is reversible. This means that the products of the reaction can react to make the original reactants. For example, ammonia and hydrogen chloride react to make solid ammonium chloride. But at the same time, some of the newly-made ammonium chloride decomposes to make ammonia and hydrogen chloride again. You can represent **reversible reactions** like this:

ammonium chloride $\rightleftharpoons$ ammonia + hydrogen chloride

▲ Burning lithium in air to make lithium oxide and lithium nitride

▲ The reaction of ammonia and hydrogen chloride is reversible

Yield

The **yield** of a substance is how much there is of it after a chemical reaction. Usually, there is a difference between the **actual yield** and the **maximum theoretical yield**:

- The maximum theoretical yield is the expected mass calculated from the reaction equation, using the masses of substances that react.
- The actual yield is the mass of product made.

The **percentage yield** compares these amounts. Use this formula to calculate it:

$$\text{percentage yield} = \frac{\text{actual yield}}{\text{maximum theoretical yield}} \times 100$$

Mr Merry heated a piece of sodium metal in chlorine gas. He calculated a maximum theoretical yield for sodium chloride of 10 g. But the actual yield was only 8 g. This means that the percentage yield was 80%:

$$\text{percentage yield} = \frac{8}{10} \times 100 = 80\%$$

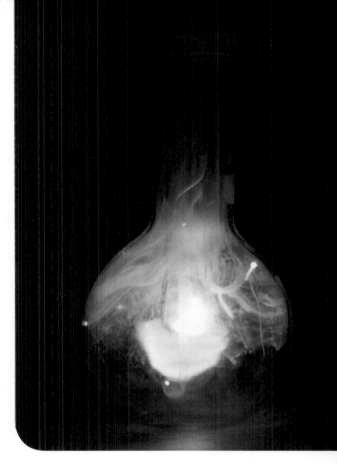

▲ The actual yield of most reactions is less than the maximum theoretical yield

Questions

1. What is a reversible reaction?

2. Sarah heats magnesium in air to make magnesium oxide. Suggest why the actual yield might be less than the maximum theoretical yield she calculated before starting.

3. Zachary mixes lead nitrate and potassium iodide solutions to make solid lead iodide. He filters the mixture to separate the solid product from the solution. Suggest why the actual yield might be less than the maximum theoretical yield he calculated before starting.

4. Grayson burns a sample of magnesium in air. He calculates that the maximum theoretical yield of magnesium oxide is 5 g, but ends up with only 4.5 g. What is the percentage yield?

Exam tip

✔ If you are asked to calculate the percentage yield in the exam, you will be given the actual yield and the theoretical yield.

Key words

reversible reaction, **yield**, **actual yield**, **maximum theoretical yield**, **percentage yield**

Course catch-up

Revision checklist

- ⭕ In covalent bonding atoms share electron pairs, forming molecules or giant covalent structures.
- ⭕ In ionic bonding atoms transfer electrons, forming ions. Positively and negatively-charged ions attract each other in giant ionic lattices.
- ⭕ Ionic compounds form when metals react with non-metals.
- ⭕ Dot and cross diagrams show the arrangement of electrons in ionic and covalent substances.
- ⭕ Ions in ionic compounds and atoms in covalent compounds have stable electronic structures.
- ⭕ In metallic bonding, layers of closely-packed ions are surrounded by a sea of delocalised electrons.
- ⭕ Substances made up of simple molecules have low melting and boiling points and do not conduct electricity. Ionic compounds have high melting points and conduct electricity when molten or in solution.
- ⭕ Diamond and graphite are forms of carbon with giant covalent structures. Fullerenes are another form of carbon used to form nanotubes.
- ⭕ Metals conduct electricity because of delocalised electrons.
- ⭕ The properties of metals are altered by forming alloys.
- ⭕ Thermosetting polymers have cross-links between chains (unlike thermosoftening polymers). They cannot be melted.
- ⭕ The nucleus of an atom contains protons (positive) and neutrons (uncharged).
- ⭕ Atomic number is the number of protons in the nucleus of an atom. Mass number is the total number of protons and neutrons in the nucleus of an atom.
- ⭕ Isotopes have the same atomic number but different mass numbers.
- ⭕ The relative formula mass of a substance, in grams, is called one mole.
- ⭕ Gas chromatography and mass spectrometry are instrumental techniques used to separate and identify compounds.
- ⭕ Empirical formulas show the simplest ratios of the elements in a compound.
- ⭕ Percentage yields can be low because product is lost in separation, or the reaction is reversible.

ATOMIC STRUCTURE

nucleus

isotopes

same atomic number but different mass number

INSTRUMENTAL METHODS

mass spectrometry

gas chromatography

CALCULATIONS

relative molecular mass

mass of one mole

reacting masses

empirical formula

percentage yields

losses due to reversible reaction separation etc

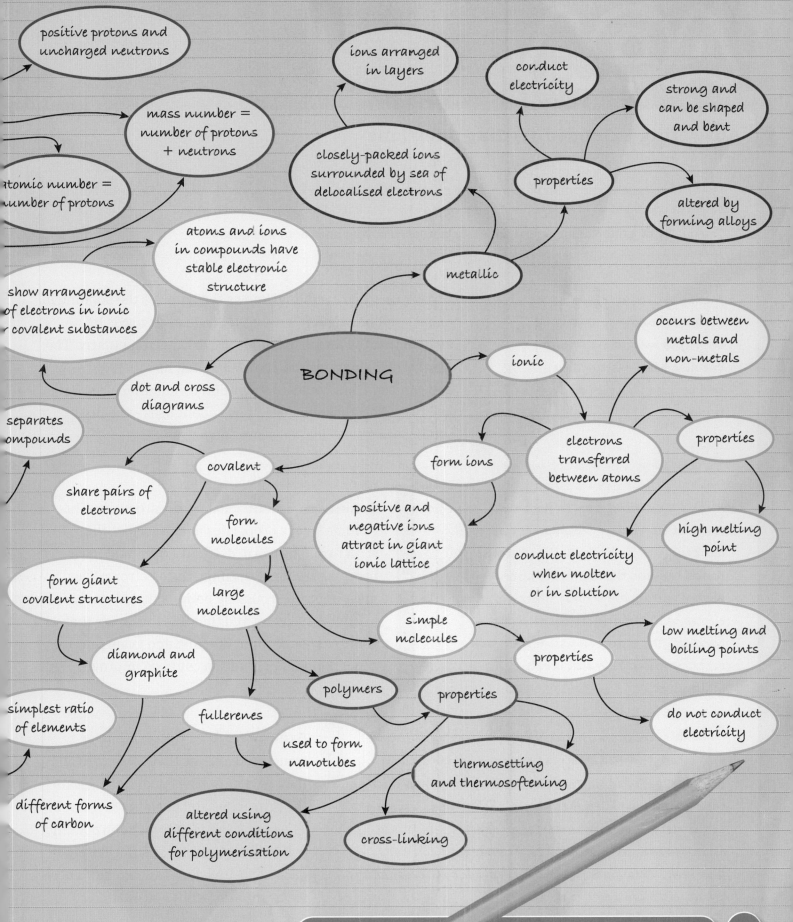

positive protons and uncharged neutrons

ions arranged in layers

conduct electricity

strong and can be shaped and bent

mass number = number of protons + neutrons

atomic number = number of protons

closely-packed ions surrounded by sea of delocalised electrons

properties

altered by forming alloys

atoms and ions in compounds have stable electronic structure

metallic

occurs between metals and non-metals

show arrangement of electrons in ionic or covalent substances

BONDING

ionic

dot and cross diagrams

electrons transferred between atoms

properties

separates compounds

covalent

form ions

high melting point

share pairs of electrons

form molecules

positive and negative ions attract in giant ionic lattice

conduct electricity when molten or in solution

form giant covalent structures

large molecules

simple molecules

low melting and boiling points

properties

simplest ratio of elements

diamond and graphite

polymers

properties

do not conduct electricity

fullerenes

used to form nanotubes

thermosetting and thermosoftening

different forms of carbon

altered using different conditions for polymerisation

cross-linking

Answering Extended Writing questions

QUESTION

Metals have several important properties that can be explained by their structure and bonding. Describe the structure and bonding in a metal, and use this to explain the important properties of metals.

The quality of written communication will be assessed in your answer to this question.

G–E

Metals are regular and have metallik bonding. This is strong so metals are strong but also bend easily. They can be used in wires and bridges.

Examiner: The candidate knows the name of the bonding in metals (but has spelt it wrongly). The answer needs to explain that the atoms in metals are arranged regularly. The answer demonstrates the knowledge that metals are strong and can bend, but could also mention conducting electricity and heat. The question doesn't ask about uses – the final sentence is irrelevant and scores no marks.

D–C

Metals are strong, can be bent and conduct electricity . Atoms are arranged close together and regularly. Metallic bonds between the atoms are strong which means metals are strong. There are electrons in the structure so metals conduct. The atoms can slide over each other making the metal bendy.

Examiner: Several good points are made. The candidate lists properties and describes the structure. The word 'delocalised' should have been used when describing the electrons. The most important missing point is that atoms are arranged in layers and that it is the layers that slide over each other. Spelling and grammar are good, though punctuation could be improved.

B–A*

In metals, the atoms are arranged regularly in layers and closely packed. They are surrounded by a sea of delocalised electrons. The force between the atoms and the electrons is called metallic bonding. This is strong so the metal has high tensyl strength The delocalised electrons can move around so it conducts electricity and heat. Metals are flexible and can be bent because the layers slide over each other.

Examiner: A very good, well-structured answer that deals with all the key points. The candidate lists all the properties and explains them well. The only error is that metallic bonding is between electrons and the closely-packed positive ions (not atoms). There is one minor spelling mistake.

Exam-style questions

1 a Ionic bonding is one of the ways in which atoms join together. Which two of the following statements about ionic bonding are true?

   i Electron pairs are shared.

   ii Molecules are formed.

   iii Positive and negative ions attract each other.

   iv It happens between metals and non-metals.

b Complete this sentence by choosing words from the list below:

Ionic compounds conduct electricity when they are _____ or _____.

solid   liquid   gas   in solution

2 99% of carbon atoms are carbon-12, with the symbol $^{12}_{6}C$. However, about 1% of carbon atoms are carbon-13, with the symbol $^{13}_{6}C$. Both have the same atomic number, but different numbers of neutrons.

a Which word is used to describe these two types of carbon atom?

b i What is the atomic number of both of these carbon atoms?

   ii How many neutrons are there in the nucleus of a carbon-13 atom?

c What can you predict about the relative atomic mass of carbon? Choose the correct answer from the list:

   i slightly less than 12

   ii exactly 12

   iii slightly more than 12

3 Ammonia, NH_3, is manufactured in industry by reacting nitrogen and hydrogen.

a Balance this equation for the manufacture of ammonia:

$$N_2 + \_\_H_2 \rightleftharpoons \_\_NH_3$$

b The predicted yield of ammonia in a reaction was 15 tonnes. However, only 6 tonnes was produced. Calculate the percentage yield.

c State three reasons why the percentage yield of this reaction is not 100%.

Extended Writing

4 Metals are very useful substances. What properties do metals have, and what uses do metals have because of them?

5 GC–MS (gas chromatography–mass spectrometry) is an instrumental method used by chemists to analyse substances. Explain why it is used and how it works.

6 The structures of diamond and graphite are shown below:

diamond   graphite

strong covalent bond   strong covalent bond

weak intermolecular forces between layers

Why do diamond and graphite have different properties?

A01 Recall the science

A02 Apply your knowledge

A03 Evaluate and analyse the evidence

Rates, energy, salts, and electrolysis

Why study rates, energy, salts, and electrolysis?

Medicines, metals, fireworks, fertilisers – you name it, chemists have helped make it. A huge variety of reaction types make useful products, including acid–base reactions and precipitation reactions. But whatever the reaction type, chemists need answers to two questions: How much energy does the reaction transfer? How fast does it go? They can then work out how to make as much product as possible as quickly as possible.

In this unit, you will find out about rates of reactions, and how to speed them up. You will learn about energy transfers in chemical reactions, too. You will discover how to make salts. Finally, you will take a look at electrolysis.

You should remember

1 Reactions happen at different speeds. Some, such as rusting, happen slowly. Others, such as explosions, happen very fast indeed.

2 Chemical reactions involve energy transfers, either from or to the surroundings.

3 Energy can be transferred in chemical reactions as heat, light, sound, or electricity.

4 Acids have a pH of less than 7, neutral solutions have a pH of 7, and the pH of an alkaline solution is greater than 7.

5 Ionic substances conduct electricity when melted or in solution.

6 Passing an electric current through an ionic compound that is melted or in solution can break down the ionic compound.

7 Metals above carbon in the reactivity series are extracted from their minerals by electrolysis.

These calcium sulfate crystals are the largest crystals known in the world. They are in the Cave of Crystals in Naica Mine, Chihuahua, Mexico. The crystals formed naturally over millions of years, and were discovered in 2000 after water was pumped out of the mine. In the lab, you can make calcium sulfate crystals from sulfuric acid and calcium oxide.

Learning objectives

After studying this topic, you should be able to:

✔ explain the meaning of the term rate of reaction

✔ use data, equations, and graphs to calculate reaction rates

Quick profit

Fernando lives near a lake in Chile. The lake has many salts dissolved in it, including lithium chloride. Fernando wants to set up a factory to make lithium carbonate tablets from the lithium chloride. The tablets treat mood disorders.

The faster Fernando can make the tablets, the sooner he will start making a profit. He experiments to find out how to maximise the **rate of reaction**. The rate of a reaction is a measure of how quickly it happens.

Fast or slow?

Some reactions, such as a firework exploding, happen in less than a second. Others, such as a metal rusting, may continue for weeks, months, or years.

▲ Rusting reactions happen very slowly

▲ Firework reactions happen quickly

A Explain why chemists need to find out about rates of reaction.

B Give two examples of very fast reactions.

Following reactions

You cannot find the rate of a reaction from its equation. You need to do experiments to discover how quickly products are made or reactants are used up.

Corinna wants to measure the rate of the reaction of calcium carbonate with hydrochloric acid. The equation for the reaction is:

calcium carbonate	+	hydrochloric acid	→	calcium chloride	+	water	+	carbon dioxide
$CaCO_3$	+	$2HCl$	→	$CaCl_2$	+	H_2O	+	CO_2

Corinna decides to measure the volume of carbon dioxide gas made as the reaction happens. She sets up the apparatus as here on the left.

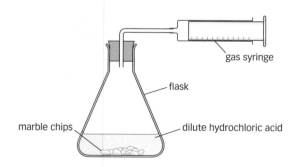

gas syringe

flask

marble chips

dilute hydrochloric acid

The reaction starts as soon as both reactants are in the flask. As carbon dioxide gas forms, it pushes out the syringe plunger. Every 30 seconds, for 240 seconds, Corinna reads the total volume of gas that has been produced up until that time.

Calculating reaction rates

Corinna's first few results are in the table. She plots all her results on a graph, too.

Corinna uses an equation to calculate the rate of reaction over the first 30 seconds:

$$\text{Reaction rate} = \frac{\text{amount of product formed}}{\text{time}} = \frac{15 \text{ cm}^3}{30 \text{ s}} = 0.5 \text{ cm}^3/\text{s}$$

Over the next 30 seconds, from 30 seconds to 60 seconds:

$$\text{Reaction rate} = \frac{(25 - 15) \text{ cm}^3}{30 \text{ s}} = \frac{10 \text{ cm}^3}{30 \text{ s}} = 0.3 \text{ cm}^3/\text{s}$$

The results of the calculations show that the reaction gets slower as time goes on.

Key words

rate of reaction

Time (s)	Volume of gas (cm³)
0	0
30	15
60	25
90	30

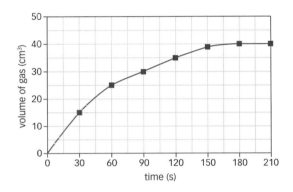

▲ The volume of gas made in the reaction of calcium carbonate and hydrochloric acid. At first the gradient is steep, showing that the rate of reaction is fast. The gradient gets less over time, showing that the reaction slows down.

Questions

1 Give two examples of very slow reactions.

2 Use results from Corinna's table to calculate the rate of the reaction between 60 seconds and 90 seconds.

3 Draw and label a diagram of the apparatus to measure the rate of reaction for the reaction below. Hydrogen is formed as a gas.

   magnesium + hydrochloric acid → magnesium chloride + hydrogen

4 Explain how Corinna's graph shows that the rate of the reaction decreases as time goes by.

5 Use the graph to answer these questions:

   (a) How much gas is made in the first 45 seconds?

   (b) How long does it take to collect 20 cm³ of gas?

   (c) Calculate the rate of reaction between 0 s and 120 s.

Exam tip

✔ Practise using the rate of reaction equation. If you are given data about amounts of reactant, you might need to use the equation below, instead of the one given above. Check your units!

$$\text{Reaction rate} = \frac{\text{amount of reactant used}}{\text{time}}$$

Learning objectives

After studying this topic, you should be able to:

✔ recall that chemical reactions happen when particles collide

✔ describe and explain the effect of changing temperature on the rate of reaction

Fast food

Chris adds chips to boiling water, at 100 °C. They cook in seven minutes. Freya adds chips of the same size to oil, at about 200 °C. They cook more quickly. The chemical reactions that happen in potatoes when they cook are quicker at higher temperatures. The rates of reaction are faster.

◀ Potatoes cook more quickly in oil than in water

▲ Reaction of sodium thiosulfate with HCl

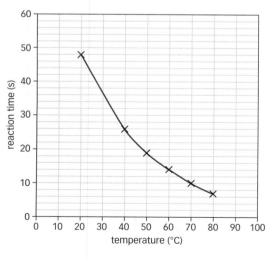

▲ Graph to show how reaction time varies with temperature

Investigating the effect of temperature

Temperature affects the rates of all reactions. Barney does an investigation to find out more. He uses this equation:

$$\text{sodium thiosulfate} + \text{hydrochloric acid} \rightarrow \text{sodium chloride} + \text{water} + \text{sulfur dioxide} + \text{sulfur}$$

$$Na_2S_2O_3 + 2HCl \rightarrow 2NaCl + H_2O + SO_2 + S$$

Barney's apparatus is shown on the left. The flask on the left shows a mixture of sodium thiosulfate and hydrochloric acid before it has reacted. The flask on the right shows the mixture of products.

Sulfur is insoluble in water, so it makes the reaction mixture turn cloudy. Barney draws a cross on a piece of paper and times how long it takes for the mixture to become so cloudy that he can no longer see the cross. He repeats the experiment at different temperatures.

The graph shows Barney's results. As temperature increases, the reaction happens more and more quickly. The reaction time gets less.

A Name the independent and dependent variables in the investigation.

B Identify the control variables in the investigation.

C Describe what happens to the rate of reaction as temperature increases.

Key words

collide, activation energy, successful collision

When particles collide

Reactions can only happen when reacting particles **collide** with each other. The colliding particles need enough energy, too. The minimum amount of energy that particles need in order to react is the **activation energy**. If two colliding particles have less energy than the activation energy, they will not react.

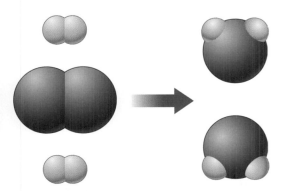

▲ Hydrogen and oxygen molecules can only react to produce water if the molecules collide

More about particles and rates

The rate of a reaction depends on
- the amount of energy transferred during collisions
- the frequency of collisions.

At low temperatures, particles move relatively slowly. So they do not collide very often. And when they do collide, the amount of energy transferred may not be enough for the particles to react.

At higher temperatures, the particles move faster. They collide more frequently. And, when the particles collide, there is more chance of a **successful collision** (one that leads to a reaction) because faster moving particles transfer more energy when they collide.

Exam tip

✔ Remember, faster moving particles collide more frequently and with more energy. So as the reaction rate increases, the reaction time gets less.

Questions

1 What happens to the rate of a reaction as the temperature decreases?

2 Suggest why the rate of production of water was not measured in the investigation.

↓ E

3 Explain, in terms of particles, what happens to the rate of reaction when the temperature is increased.

↓ C

4 Use the graph to answer these questions:

(a) Explain, in terms of particles, why the rate of reaction doubles between 60 °C and 80 °C.

↓ A*

(b) Estimate the rate of reaction at 30 °C.

Learning objectives

After studying this topic, you should be able to:

✔ describe and explain the effect of changing concentration on rates of reaction

✔ describe and explain the effect of changing pressure on rates of reaction

Faster and faster

You can speed up reactions by increasing the temperature of the reactants. But this is not the only way to make reactions happen faster. You can also speed up reactions by

- increasing the concentration of reactants in solution
- increasing the pressure of reacting gases
- increasing the surface area of solid reactants
- using a catalyst.

Rate and concentration

Earl investigates the effect of concentration on reaction rate. He sets up this apparatus.

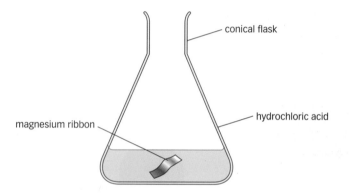

He measures the time for the magnesium to disappear when it reacts with hydrochloric acid of different concentrations. For each test, he uses the same length of magnesium ribbon, and the same volume and temperature of acid.

Earl draws a graph of his results.

A Name the independent and dependent variables in Earl's investigation.

B List three control variables for the investigation.

C What happens to the reaction time as the acid concentration increases?

D What happens to the rate of reaction as the acid concentration increases?

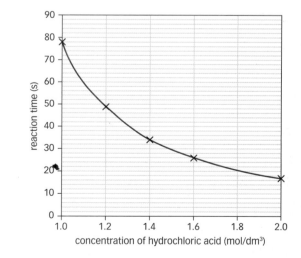

▲ Graph to show how reaction time varies with concentration

Particles and concentration

The concentration of a solution is a measure of how much solute is dissolved in the solvent. The more concentrated a solution, the greater the number of solute particles that are dissolved in a certain volume of solvent.

If this represents a 1 mol/dm³ solution of acid ...

then this represents a 2 mol/dm³ solution of the same acid.

▲ There are double the number of acid particles in the same volume of water.

In Earl's investigation, as the acid concentration increased, so the acid particles became more crowded. The frequency of collisions between the acid particles and the magnesium particles increased, so the rate of reaction increased.

Rate and pressure

Just a tiny spark can set fire to a mixture of hydrogen and oxygen.

$$\text{hydrogen} + \text{oxygen} \rightarrow \text{water}$$
$$2H_2 + O_2 \rightarrow 2H_2O$$

The two gases mix together completely, so their particles collide frequently. If you increase the pressure of the gas mixture, their particles become more crowded. They now collide even more frequently, so the rate of reaction increases.

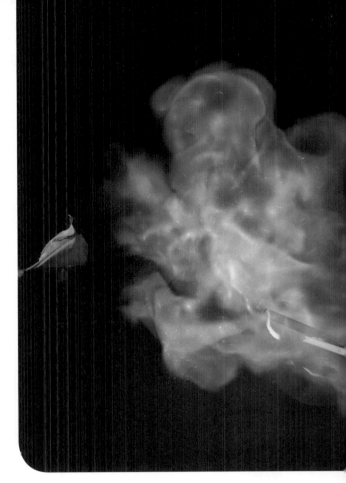

▲ Hydrogen burns explosively in air

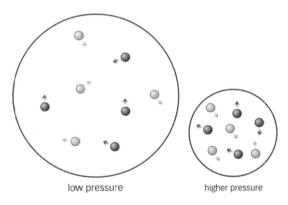

low pressure   higher pressure

▲ Gas particles collide more frequently at higher pressures

Questions

1 Describe what happens to the rate of reaction involving gases when the pressure is decreased.

2 Describe what happens to the rate of a reaction involving solutions when the concentration is reduced.

↓ E

3 In Earl's investigation, what was the relationship between reaction time and acid concentration?

↓ C

4 Explain, in terms of particles, why the rate of reaction increases if the pressure or concentration of the reactants is increased.

↓ A*

Exam tip   AQA

✓ Increasing the concentration of reactants in solutions, or the pressure of reacting gases, increases the frequency of collisions and so increases the rate of reaction.

Learning objectives

After studying this topic, you should be able to:

✔ describe and explain the effect of changing surface area on rates of reaction

Bang!

TNT (trinitrotoluene) is an explosive. It reacts very fast to release a huge volume of hot gases in a short time. The expanding gases create a shock wave that travels very fast indeed. The shock wave damages objects in its path.

▲ Explosives are used to break the rock face in open cast mines

Powders like flour and custard can cause explosions, too. They burn very easily. This makes them dangerous in factories where they are made or used. The factories have strict safety rules to stop dust escaping into the air inside buildings, and to prevent sparks or naked flames.

Explaining rate and surface area

A reaction involving a powder happens faster than a reaction involving a lump of the same reactant. A powder has a bigger **surface area** than a lump of the same mass. This is because particles that were inside the lump become exposed on the surface when it is crushed.

Reactant particles must collide for a reaction to happen. The larger the surface area, the greater the frequency of collisions, and so the faster the reaction.

▲ Powdered milk burns explosively

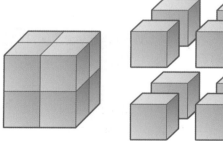

▲ Eight cubes with sides of 1 cm length have the same volume as 1 cube with sides 2 cm length. But the 1 cm cubes have twice the surface area.

Investigating rate and surface area

Emily collects data to find out more about how surface area affects reaction rate. She uses this apparatus.

▲ Reaction of calcium carbonate with HCl

The equation for the reaction in the investigation is:

$$\text{calcium carbonate} + \text{hydrochloric acid} \rightarrow \text{calcium chloride} + \text{water} + \text{carbon dioxide}$$

$$CaCO_3 + 2HCl \rightarrow CaCl_2 + H_2O + CO_2$$

- First, Emily adds a lump of calcium carbonate to excess hydrochloric acid. Every minute, she measures the loss of mass as carbon dioxide gas escapes from the reacting mixture.
- Next, Emily adds powdered calcium carbonate with the same mass as the lump to excess hydrochloric acid. Again, she measures the loss of mass as carbon dioxide escapes.

Emily plots her results on a graph.

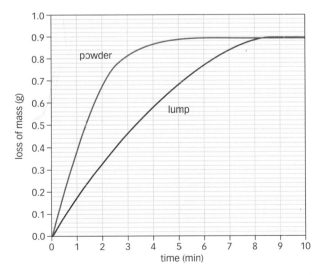

▲ The loss in mass during the reaction of calcium carbonate with hydrochloric acid

A Identify the dependent, the independent, and the control variables in the investigation.

B Explain how the graph shows that the powder reacts faster than the lumps.

Exam tip

✔ Increasing the surface area of solid reactants increases the frequency of collisions and so increases the rate of reaction.

Questions

1 In general, which react faster, lumps or powders?

2 Explain why increasing the surface area of a solid reactant increases the rate of a reaction.

3 Use the graph to answer these questions:

  (a) When did each of the reactions finish?

  (b) Which reaction was faster during the first two minutes?

4 Calculate the mean (average) rate of reaction for each reaction shown in the graph.

5 Explain why powders are dangerous in factories.

Learning objectives

After studying this topic, you should be able to:

- ✓ explain what catalysts are and what they do
- ✓ explain why catalysts are important in industry

Cats in cars

There are millions of cars in the UK. And each one is a source of pollution. To tackle this problem, car exhaust systems are fitted with catalytic converters, or 'cats'. In a catalytic converter, chemical reactions convert harmful exhaust gases into less harmful ones. For example:

$$\text{carbon monoxide} + \text{oxygen} \rightarrow \text{carbon dioxide}$$
$$2CO + O_2 \rightarrow 2CO_2$$

This reaction can happen on its own, but under the conditions in an exhaust system the reaction is too slow to get rid of carbon monoxide before it goes into the air. The catalytic converter speeds up the reaction. But how? What's inside a cat?

Catalysts

You can speed up reactions by increasing
- the temperature
- the concentration, if solutions are involved
- the pressure, if gases are involved
- the surface area, if solids are involved.

You can also use a **catalyst** to make a reaction faster. A catalyst is a substance that increases the rate of a chemical reaction without being used up in the reaction. So if you add 1 g of a catalyst to a reaction mixture, there is still 1 g of it left when the reaction has finished.

Catalytic converters contain two or three precious metals as catalysts, usually chosen from platinum, rhodium, and palladium.

Just a little

In July 2010, just one gram of platinum cost about £32. Luckily, a small amount of catalyst will **catalyse** the reaction between large amounts of reactants, so a typical catalytic converter needs only 3 g of platinum catalyst.

Catalysts are specific to particular reactions. A substance that acts as a catalyst for one reaction may not work as a catalyst for another reaction.

▲ Catalytic converters in cars convert harmful exhaust gases into less harmful ones

A What is a catalyst?

B Describe one way in which platinum is used as a catalyst.

Catalysts in the lab

If you've ever bleached your hair, you've probably used hydrogen peroxide solution.

Hydrogen peroxide is usually diluted in water, where it breaks down *very* slowly:

hydrogen peroxide $\rightarrow$ water $+$ oxygen

$$2H_2O_2 \rightarrow 2H_2O + O_2$$

Powdered manganese(IV) oxide, MnO_2, catalyses this decomposition reaction. If you add a small amount of manganese(IV) oxide to hydrogen peroxide solution, lots of oxygen quickly forms.

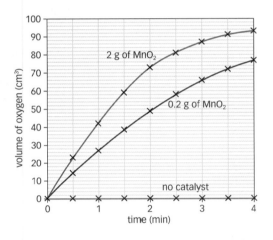

◀ Adding a lot more catalyst makes little difference to the rate at which hydrogen peroxide breaks down

▲ Hydrogen peroxide decomposes when a manganese(IV) oxide catalyst is added

Questions

1 Give an example of a catalyst and the reaction it catalyses.

2 Explain why catalysts are important in industry.

3 Use the graph to answer these questions:

(a) What was the effect on the rate of reaction of adding manganese(IV) oxide?

(b) Explain which reaction was faster in the first two minutes.

(c) How can you tell that the reactions were not complete after four minutes?

4 Calculate the mean rate of reaction for each reaction shown in the graph.

5 Suggest why adding ten times as much manganese(IV) oxide did not increase the rate of reaction by ten times.

Key words

catalyst, catalyse

Exam tip

✔ Catalysts are useful in the chemical industry. Without them, many industrial processes would be too slow to be profitable. For example, iron is used in the Haber process, which makes ammonia for fertilisers and explosives. The iron catalyst increases the rate of the reaction between nitrogen and hydrogen. So more ammonia is made in a shorter time.

23: Energy and chemical reactions

Learning objectives

After studying this topic, you should be able to:

✔ explain what exothermic reactions are, and give examples of them

Key words

exothermic reaction

Firework fun

Fabulous fireworks light up the sky at Diwali and the new year. Chemical reactions in fireworks transfer energy to the surroundings as heat, light, and sound. Reactions that transfer energy to the surroundings are called **exothermic reactions**.

▲ Firework display

A What is an exothermic reaction?

B State three forms of energy that can be transferred by chemical reactions.

Combustion reactions

Exothermic reactions do not happen only in fireworks. All combustion reactions transfer energy to the surroundings, mainly as heat and light. The heat they transfer is useful for cooking, heating homes, and generating electricity. Power stations burn coal to heat water into steam. The steam turns turbines which generate electricity.

More exothermic reactions

Neutralisation reactions are exothermic, too. Daisha adds 50 cm³ of dilute hydrochloric acid to 50 cm³ of sodium hydroxide solution. She records the temperatures before and after the reaction.

Substance	Temperature (°C)
Hydrochloric acid, before mixing	20
Sodium hydroxide solution, before mixing	20
Reaction mixture, immediately after mixing	49
Reaction mixture, 1 hour after mixing	20

Hydrochloric acid and sodium hydroxide react in a neutralisation reaction:

$$\text{hydrochloric acid} + \text{sodium hydroxide} \rightarrow \text{sodium chloride} + \text{water}$$

$$HCl + NaOH \rightarrow NaCl + H_2O$$

The reaction gives out energy. At first, this energy heats up the reacting mixture. Then heat is transferred from the mixture to the surroundings, and the mixture cools to room temperature.

Some other types of reaction are exothermic, including:

- many oxidation reactions – for example the reaction in which potassium chlorate oxidises the glucose sugar in a jelly baby, causing it to burst into flames and make a screaming sound

◀ Jelly babies

- displacement reactions – for example when zinc reacts with copper sulfate solution to make copper and zinc sulfate.

Using exothermic reactions

Hand warmers use exothermic changes to produce heat. In one type, iron is quickly oxidised when you activate the hand warmer. The reaction is exothermic, so it transfers heat to your hands.

Self-heating coffee cans have two compartments. One contains cold coffee. The other contains reactants which react together in an exothermic reaction. This transfers heat to the coffee. Delicious!

▲ Glow worm at night

Did you know...?

Chemical reactions in glow worms transfer light energy to the surroundings.

Questions

1 Name three types of reaction which are exothermic.

2 Give two examples of exothermic reactions which are useful.

↓ E

3 Describe what happens to the temperature of a reaction mixture during and after an exothermic reaction.

↓ C

4 Suggest how you could find out whether the reaction of magnesium with dilute hydrochloric acid is exothermic or not. What results would you expect if the reaction is exothermic?

↓ A*

5 Explain how Daisha's data show that the neutralisation reaction is exothermic.

Learning objectives

After studying this topic, you should be able to:

- ✔ recall that, in chemical reactions, energy can be transferred from or to the surroundings
- ✔ explain what endothermic reactions are, and give examples of them
- ✔ recall that if a reaction is exothermic in one direction it is endothermic in the opposite direction

---

A What is an endothermic reaction?

B Give two examples of endothermic reactions.

Sherbet fizz

What happens when you eat a sherbet? It fizzes in your mouth, and your tongue feels cold. The fizzing happens when two sherbet ingredients – sodium hydrogencarbonate and citric acid – react together. This is an **endothermic reaction**. It takes in energy from the surroundings, in this case in the form of heat from your tongue.

▲ Sherbet sweets

The sherbet reaction also happens if you put the sherbet in water. You can tell it is endothermic because the temperature decreases.

More endothermic reactions

Here are some more examples of endothermic processes:

▲ When ammonium nitrate dissolves in water, the temperature falls so much that a drop of water can freeze a container to a block of wood

▲ Energy must be supplied to get thermal decomposition reactions to happen. Here, copper carbonate is decomposing to make copper oxide and carbon dioxide.

Using endothermic reactions

Sports injury packs relieve pain on the football field and hockey pitch. Some are based on an endothermic reaction – activating them starts off a chemical reaction that takes in heat energy from the injured leg or arm.

Reversible reactions

All chemical reactions involve energy transfers. If a reversible reaction is exothermic in one direction, it is endothermic in the opposite direction. The same amount of energy is transferred in each direction:

- Pawel takes some blue copper sulfate crystals. The crystals are **hydrated** – they contain water. The formula of hydrated copper sulfate is $CuSO_4.5H_2O$. Pawel heats the crystals. A white powder forms. The white powder is **anhydrous** copper sulfate, $CuSO_4$. It has no water in it. This process is endothermic – it takes in heat energy from the surroundings.
- Pawel waits for the white powder to cool. Then he adds a few drops of water to it. Blue crystals form again, and heat energy is given out. The process is exothermic.

You can summarise Pawel's reactions like this:

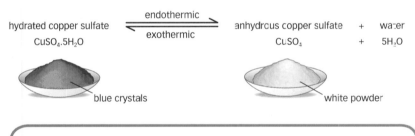

hydrated copper sulfate $\underset{\text{exothermic}}{\overset{\text{endothermic}}{\rightleftharpoons}}$ anhydrous copper sulfate + water
$CuSO_4.5H_2O$ $CuSO_4$ + $5H_2O$

blue crystals — white powder

▲ Sports injury packs cool damaged muscles

Key words

endothermic reaction, hydrated, anhydrous

Questions

1 Give an example of a useful endothermic reaction.

2 Draw up a table to summarise the differences between endothermic and exothermic reactions. Include ideas about temperature changes and energy transfer.

3 Draw diagrams to summarise what Pawel did in his experiment, and to show the direction in which the reaction is exothermic, and the direction in which it is endothermic.

Exam tip **AQA**

✔ Think of a fire exit sign to help you remember which way round exothermic and endothermic reactions go. A fire is hot and you go out of an exit. Exothermic reactions get hot as they give energy out.

Learning objectives

After studying this topic, you should be able to:

✔ explain what makes solutions acidic or alkaline

✔ explain what happens in neutralisation reactions

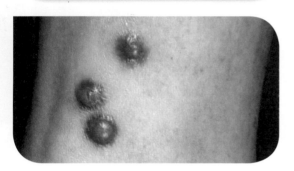

▲ Ant stings are acidic, like bee stings

Did you know...?

Ant stings are acidic, like bee stings, but wasp stings are alkaline.

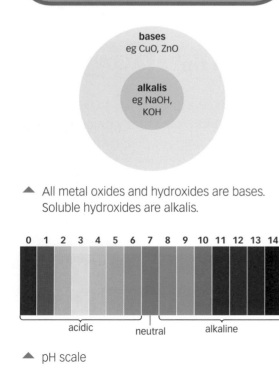

▲ All metal oxides and hydroxides are bases. Soluble hydroxides are alkalis.

| 0 | 1 | 2 | 3 | 4 | 5 | 6 | 7 | 8 | 9 | 10 | 11 | 12 | 13 | 14 |

acidic   neutral   alkaline

▲ pH scale

Bee sting

Harry gets stung by a bee. It hurts! He puts toothpaste on the sting. Soon, the sting is less painful. The toothpaste has neutralised – 'cancelled out' – the acid in the bee sting.

▲ Bee stings are acidic

Alkalis and bases

Toothpaste is not the only substance that neutralises acids. All bases and alkalis do:

• **Bases** are metal oxides and hydroxides, for example, copper oxide.

• Hydroxides that are soluble are called **alkalis**. Sodium hydroxide and potassium hydroxide are alkalis.

Charged particles called **hydroxide ions**, OH^-(aq), make solutions alkaline. The (aq) shows that the hydroxide ion is dissolved in water.

The pH scale

The **pH scale** measures the acidity or alkalinity of a solution:

• A solution of pH 7 is neutral.

• A solution with a pH of less than 7 is acidic.

• A solution with a pH of more than 7 is alkaline.

You can use indicators such as Universal indicator to show the pH of a solution.

A Give the formula of the particle that makes solutions alkaline.

B A solution has a pH of 6.8. Is the solution acidic, alkaline, or neutral?

Key words

base, alkali, hydroxide ion, pH scale, hydrogen ion, neutralisation reaction

Neutralisation

Hydroxide ions make solutions alkaline. But what makes solutions acidic? The answer is **hydrogen ions**, $H^+(aq)$. Hydrogen ions only make solutions acidic if they are dissolved in water.

When sodium hydroxide neutralises hydrochloric acid, the products are sodium chloride and water:

hydrochloric acid + sodium hydroxide → sodium chloride + water

$$HCl(aq) + NaOH(aq) \rightarrow NaCl(aq) + H_2O(l)$$

In this reaction, hydrogen ions from the hydrochloric acid react with hydroxide ions from the sodium hydroxide to produce water. You can represent the reaction – and all **neutralisation reactions** – like this:

$$H^+(aq) + OH^-(aq) \rightarrow H_2O(l)$$

The (l) shows that the water is liquid. Sodium ions and chloride ions play little part in the reaction of hydrochloric acid with sodium hydroxide.

Amazing ammonia

Like metal hydroxides, ammonia dissolves in water to produce an alkaline solution. The formula of ammonia is NH_3. In water, it forms hydroxide ions like this:

ammonia(g) + water(l) → ammonium hydroxide(aq)

$$NH_3(g) + H_2O(l) \rightarrow NH_4OH(aq)$$

When dissolved in water, the ammonium and hydroxide ions are separated and surrounded by water molecules. So ammonium hydroxide solution contains $NH_4^+(aq)$ ions and $OH^-(aq)$ ions. The $OH^-(aq)$ ions make the solution alkaline.

Exam tip

✔ Learn the neutralisation equation,
$H^+(aq) + OH^-(aq) \rightarrow H_2O(l)$

Questions

1 Give the formula of the ion that makes solutions acidic.

2 A solution has a pH of 11.2. Is it acidic, alkaline, or neutral?

3 Describe the difference between a base and an alkali.

4 Explain the meaning of the symbols (l) and (aq).

5 Explain the meaning of the neutralisation equation: $H^+(aq) + OH^-(aq) \rightarrow H_2O(l)$.

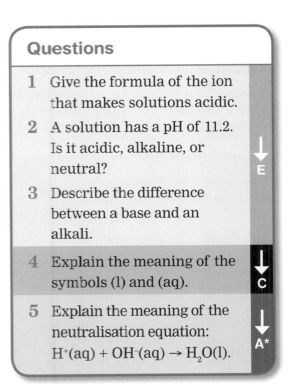

Learning objectives

After studying this topic, you should be able to:

- ✔ explain what a salt is
- ✔ describe how to make soluble salts from acids and metal oxides

▲ *Seizure*, 2008. To create this art installation, the artist allowed copper sulfate crystals to grow in a London flat.

Did you know...?

You can use copper sulfate to treat aquarium fish for parasitic infections. But don't add too much – copper ions are toxic to fish!

Key words

salt, soluble, crystallisation

Seizure

In 2008, artist Roger Hiorns poured 75 000 litres of copper sulfate solution through a hole in the ceiling of an abandoned London flat. Crystals began to grow. A few weeks later, Roger pumped out the remaining solution. The inside of the flat was covered in stunning jewel-like crystals, glinting and other-worldly. Roger named his piece of art 'Seizure'.

Salts

Copper sulfate is an example of a **salt**. Of course it is different from the substance we normally call salt – that's sodium chloride. And you certainly can't eat copper sulfate – swallowing just 1 g would make you vomit and turn yellow.

A salt is a compound that contains metal ions and that can be made from an acid.

Which acid?

Different acids make different types of salt:
- Hydrochloric acid (HCl) makes chlorides, for example sodium chloride.
- Sulfuric acid (H_2SO_4) makes sulfates, for example copper sulfate.
- Nitric acid (HNO_3) makes nitrates, for example zinc nitrate.

Metal ions for salts

Several types of substances can supply metal ions to salts, including:
- metals, such as magnesium
- insoluble bases, such as copper oxide
- alkalis, such as sodium hydroxide.

A What is a salt?

B Suggest which metal and acid you could use to make magnesium chloride.

Making a soluble salt from an insoluble base

Copper sulfate dissolves in water, so chemists call it a **soluble** salt. You can make copper sulfate by reacting sulfuric acid with copper oxide, an insoluble base:

copper oxide + sulfuric acid → copper sulfate + water

$$CuO(s) \quad + \quad H_2SO_4(aq) \quad \rightarrow \quad CuSO_4(aq) \quad + \quad H_2O(l)$$

The (s) in the symbol equation means that the copper oxide is solid.

Here's how to make the copper sulfate:

Step 1 Add a little copper oxide to a sample of dilute sulfuric acid. Keep adding copper oxide until no more reacts. There will now be a blue solution of copper sulfate mixed with some black copper oxide powder that has not reacted.

Step 2 Filter to remove the unreacted copper oxide powder.

Step 3 Heat the blue solution over a water bath, until about half its water has evaporated.

Step 4 Leave the solution to stand for a few days. Crystals will slowly form. This is **crystallisation**.

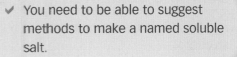

✔ You need to be able to suggest methods to make a named soluble salt.

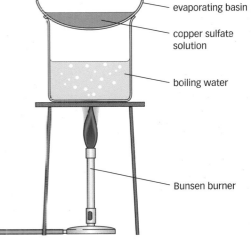

③
- evaporating basin
- copper sulfate solution
- boiling water
- Bunsen burner

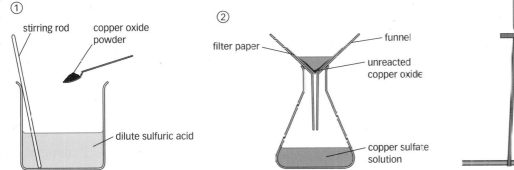

① stirring rod — copper oxide powder — dilute sulfuric acid

② filter paper — funnel — unreacted copper oxide — copper sulfate solution

Questions

1 Name the type of salt formed from nitric acid.

2 Explain the meaning of the word crystallisation.

3 Suggest which insoluble base (metal oxide) and which acid you could use to make the salts listed below. Write word equations for the reactions:

   (a) zinc nitrate   (b) copper chloride   (c) magnesium sulfate.

4 Describe how to make zinc chloride from zinc oxide and hydrochloric acid. Include a balanced equation for the reaction. Useful formulae: ZnO, $ZnCl_2$.

↓ E

↓ C

↓ A*

Learning objectives

After studying this topic, you should be able to:

- ✔ describe how to make soluble salts from acids and metals, and acids and alkalis
- ✔ suggest methods to make a named soluble salt

Magnesium sulfate

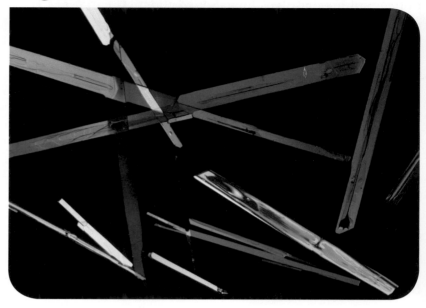

▲ Magnesium sulfate crystals

Magnesium sulfate forms beautiful crystals. It is useful, too. It helps potato, rose, and tomato plants grow well. It is also an effective laxative.

Magnesium sulfate occurs naturally in the Earth's crust. You can also make the salt in the laboratory.

Making a soluble salt from a metal

Magnesium sulfate forms in the reaction of magnesium with sulfuric acid:

magnesium + sulfuric acid → magnesium sulfate + hydrogen

$$Mg(s) + H_2SO_4(aq) \rightarrow MgSO_4(aq) + H_2(g)$$

The (g) shows that hydrogen is a gas.

> **A** Give the meanings of the symbols (s), (l), (g), and (aq).

Here's how to make the magnesium sulfate:

- Add a small piece of magnesium ribbon to a sample of dilute sulfuric acid. Keep on adding magnesium until the bubbling stops and there is a little solid magnesium in the colourless solution of magnesium sulfate.
- Filter to remove unwanted magnesium ribbon.
- Heat the colourless solution over a water bath, until about half its water has evaporated.
- Allow the solution to crystallise by leaving it to stand for a few days.

Not all metals react with acids to form salts. Some, like copper, are not reactive enough. Other metals, for example sodium, are too reactive – it is dangerous to add sodium metal to dilute acid in the laboratory.

> **B** Suggest which metal and acid you could use to make zinc nitrate.

Making a soluble salt from an alkali

Sodium chloride can be made in the lab by burning sodium in chlorine gas. But it is safer and simpler to make this salt by reacting dilute hydrochloric acid with sodium hydroxide solution:

$$\text{sodium hydroxide} + \text{hydrochloric acid} \rightarrow \text{sodium chloride} + \text{water}$$

$$NaOH(aq) + HCl(aq) \rightarrow NaCl(aq) + H_2O(l)$$

The pictures below show how to make the sodium chloride.

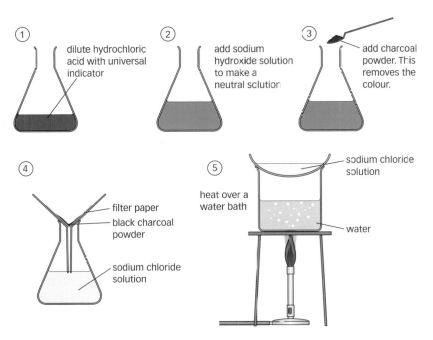

▲ Instructions for making sodium chloride

Questions

1 Name the salt made when you react magnesium with sulfuric acid.

2 When making magnesium chloride from a metal and an acid, how can you tell when you have added enough magnesium to the hydrochloric acid?

3 When making potassium chloride from an acid and alkali, how can you tell when you have added enough potassium hydroxide to the hydrochloric acid?

4 Suggest which alkali and which acid you could use to make the salts listed below. Write word equations for the reactions:
   (a) potassium nitrate
   (b) sodium nitrate.

5 Describe how to make magnesium chloride ($MgCl_2$) from a metal and an acid. Include a balanced equation for the reaction.

Learning objectives

After studying this topic, you should be able to:

- ✔ suggest the substances needed to make an insoluble salt
- ✔ give examples of how precipitation reactions are useful

Key words

precipitate, precipitation reaction, ionic equation

Exam tip AQA

- ✔ You need to be able to suggest pairs of solutions that react to make a named insoluble salt in a precipitation reaction.

A What is a precipitation reaction?

Chemical magic?

Lead nitrate solution is colourless and transparent. So is potassium iodide solution. But if you mix the solutions together you make something quite different – a bright yellow **precipitate**. A precipitate is a suspension of small solid particles, spread throughout a liquid or solution. It makes the mixture look cloudy.

◀ Lead iodide forms as a precipate if you mix lead nitrate and potassium iodide solutions

If you filter the yellow mixture, a yellow solid collects on the filter paper. A colourless solution drips into the flask.

Explaining precipitation

The equation for the reaction that makes the yellow precipitate is:

lead nitrate + potassium iodide → lead iodide + potassium nitrate

$$Pb(NO_3)_2(aq) + 2KI(aq) \rightarrow PbI_2(s) + 2KNO_3(aq)$$

The symbol equation shows that the lead iodide forms as a solid. It is insoluble in water. So the yellow precipitate is lead iodide. The reaction is a **precipitation reaction**, since it forms a precipitate.

The potassium and nitrate ions do not change in the reaction. You can use an **ionic equation** to summarise the reaction. It shows only the ions that take part.

$$Pb^{2+}(aq) + 2I^-(aq) \rightarrow PbI_2(s)$$

Making other insoluble salts

You can make other insoluble salts in precipitation reactions, too. For example:

| barium chloride | + | sodium sulfate | → | barium sulfate | + | sodium chloride |

$$BaCl_2(aq) + Na_2SO_4(aq) \rightarrow BaSO_4(s) + 2NaCl(aq)$$

Using precipitation reactions

Precipitation reactions are useful for removing unwanted ions from solution. For example, waste water may contain poisonous lead, manganese, or chromium ions. These can be removed by adding a solution that contains negative ions that form a precipitate with the metal ions. For example:

$$Cr^{3+}(aq) + 3OH^-(aq) \rightarrow Cr(OH)_3(s)$$

The hydroxide ions are supplied by calcium hydroxide solution. Filtering removes the chromium(III) hydroxide precipitate.

Working out formulae

If you know the charges of the ions in a compound, you can work out its formula. Overall, ionic compounds have no electrical charge. The positive and negative charges cancel each other out. So formulae must show equal numbers of positive and negative charges. For example, to find the formula of barium chloride:

- The charges on the ions are Ba^{2+} and Cl^-.
- A neutral compound needs two Cl^- ions for every one Ba^{2+} ion.
- The formula of the compound is $BaCl_2$.

The table shows the charges on some ions.

Charge	Examples of ions with this charge			
−2	O^{2-}	S^{2-}	SO_4^{2-}	
−1	Cl^-	Br^-	I^-	NO_3^-
+1	Li^+	Na^+	K^+	
+2	Mg^{2+}	Ca^{2+}	Ba^{2+}	
+3	Al^{3+}			

▲ Barium sulfate precipitate

B Name two solutions that react to make a precipitate of barium sulfate.

Questions

1 What is a precipitate?

2 Write a word equation for the formation of lead iodide from potassium iodide and lead nitrate solutions.

↓ E

3 Write down the formulae of magnesium chloride, potassium sulfate, and aluminium iodide.

↓ C

4 Suggest two solutions you could mix to make a precipitate of copper hydroxide.

5 Suggest a solution you could mix with silver nitrate solution to make a precipitate of silver chloride.

↓ A*

Learning objectives

After studying this topic, you should be able to:

✔ describe what happens at the electrodes in electrolysis

✔ use half equations to represent reactions at electrodes

Key words

electrolysis, electrolyte, reduction, oxidation, half equation

Extraordinary element

What do the objects in the pictures have in common?

All use the element lithium, or its compounds. Many electronic goods are powered by lithium batteries. Lithium stearate grease is a useful vehicle lubricant. Submarines use lithium peroxide to remove carbon dioxide from the air.

You can't just dig lithium out of the ground. It is much too reactive to exist on its own. Most lithium is produced by the **electrolysis** of lithium salts.

What is electrolysis?

When an ionic compound is melted, or dissolved in water, its ions are free to move about within the liquid or solution. Passing an electric current through the liquid or solution breaks down the ionic compound into simpler substances. This is electrolysis. The solution of the substance that is broken down is called the **electrolyte**.

During the electrolysis of molten (melted) lead bromide:
- Positive lead ions move towards the negative electrode.
- Negative bromide ions move towards the positive electrode.

negative electrode ⊖   ⊕ positive electrode

Key:
- ● lead ion, Pb²⁺
- ● bromide ion, Br⁻

Note: This is a simplified diagram. In fact, the whole liquid is made up of lead ions and bromide ions only.

▲ Electrolysis of molten lead bromide

A What is electrolysis?

B In electrolysis, what type of ion moves towards the negative electrode?

What happens at the electrodes?

- At the negative electrode, positively charged ions gain electrons. This is **reduction**. In the example, lead ions gain electrons. They are reduced.
- At the positive electrode, negatively charged ions lose electrons. This is **oxidation**. In the example, bromide ions lose electrons. They are oxidised.

Oxidation is not just about adding oxygen – a better definition is that oxidation is the loss of electrons.

Did you know...?

The metals aluminium, sodium, potassium, and magnesium are all produced by electrolysis. Hydrogen fuel can be made from the electrolysis of water.

Chemists use **half equations** to show what happens at the electrodes during electrolysis. Electrons are represented by e^-. You can balance a half equation by adding or subtracting electrons until the total charge on each side is equal.

For example, in the electrolysis of lead bromide:
- At the negative electrode, $Pb^{2+} + 2e^- \rightarrow Pb$
- At the positive electrode, $2Br^- - 2e^- \rightarrow Br_2$
  This is also written as $2Br^- \rightarrow Br_2 + 2e^-$.

The positive electrode half equation also shows that, at this electrode, bromide ions lose electrons. The resulting atoms join together in pairs to form bromine molecules, Br_2.

Questions

1 In the electrolysis of lead bromide, which ion moves to the negative electrode?

2 Explain why electrolysis only breaks down melted or dissolved ionic compounds.

3 Give two examples of uses of electrolysis.

4 Write half equations to show what happens at the electrodes during the electrolysis of copper chloride solution. Ion formulae: Cu^{2+} and Cl^-

5 Suggest economic and environmental reasons for using renewably generated electricity for the electrolysis of water to make hydrogen fuel.

Exam tip

- ✔ Negative ions lose electrons at the positive electrode. This is oxidation.
- ✔ Positive ions gain electrons at the negative electrode. This is reduction.
- ✔ Use OIL RIG to help you remember – Oxidation Is Loss, Reduction Is Gain.

30: Electroplating

Learning objectives

After studying this topic, you should be able to:

- ✔ explain how electroplating works
- ✔ give reasons for electroplating objects

Key words

electroplating

All that glistens is not gold

Raj had his personal music player covered in gold. It looks fantastic.

▲ This personal music player is covered in gold

Linda has a shiny motorbike. It is made mainly from steel. But steel rusts easily. So during its manufacture, the bike was coated with a layer of chromium metal. Chromium resists corrosion, so it protects the bike.

▲ This motorbike has a thin layer of chromium on its surface

Food cans are made from steel. They have a very thin coating of tin. This stops the steel rusting. You couldn't make cans from tin alone – they would be too expensive.

How were these objects coated with gold, chromium, or tin? Read on to find out.

Did you know...?

You can't electroplate steel directly with chromium, because chromium doesn't stick well to steel. You have to electroplate steel first with copper, then with nickel, and finally with chromium.

Electroplating

The music player, bike, and cans were all coated by a process called **electroplating**. Electroplating happens in electrolysis cells, like the one shown on the right.

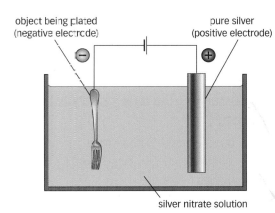

How to electroplate a fork with silver

> **A** What is electroplating?
>
> **B** Give two reasons for electroplating objects.

What happens at the electrodes?

In the electrolysis cell:

- Positive silver ions move to the negative electrode. Here, they gain electrons to form silver metal. The ions have been reduced.
- At the positive electrode, silver ions from the electrode lose electrons and go into solution. The silver has been oxidised.

You can write half equations for the reactions at the electrodes:

- Negative electrode $Ag^+ + e^- \rightarrow Ag$
- Positive electrode $Ag - e^- \rightarrow Ag^+$

Questions

1 Suggest why Raj got his music player electroplated with gold.

2 Suggest a reason for electroplating steel cutlery with silver.

3 Suggest a reason for electroplating a car bumper with chromium.

4 Explain why, in electroplating, the object to be coated must always be the negative electrode of an electrolysis cell.

5 Draw a diagram to show how you could set up a nickel electrode and a nickel sulfate solution to electroplate a paper clip with nickel.

Exam tip

✔ The object to be electroplated must always be the negative electrode of an electrolysis cell.

Learning objectives

After studying this topic, you should be able to:

✔ describe and explain how aluminium is extracted from its ore by electrolysis

▲ The properties of aluminium make it suitable for take-away food containers

Did you know…?

Aluminium used to be more highly prized than gold. Apparently, Napoleon III, Emperor of the French, gave a special dinner at which the most honoured guests had aluminium cutlery, whilst the others had to make do with gold.

▲ Modern aluminium cutlery

Awesome aluminium

There is huge demand for aluminium – more than 50 million tonnes of the metal was produced in 2006. Aluminium conducts electricity well, resists corrosion, and has a low density. These properties make it perfect for planes, packaging, and power lines. In 2010, a tonne of aluminium cost about £670, compared to about £280 for a tonne of steel.

▲ Aluminium ingots are valuable

Aluminium compounds are everywhere. Aluminium is the most abundant metal in the Earth's crust. So why is aluminium so expensive?

Extracting aluminium

Aluminium exists naturally mainly as aluminium oxide, in **bauxite** ore. Aluminium is relatively high in the reactivity series. Its oxide cannot be reduced by heating with carbon. So aluminium is extracted by electrolysis. This is what makes aluminium expensive – an aluminium plant producing about 120 000 tonnes a year needs more than 200 megawatts of electricity, enough to meet the needs of a small town.

Here's how to extract aluminium from bauxite:
- Remove impurities from the ore to get pure aluminium oxide.
- Dissolve the pure aluminium oxide in molten **cryolite**. The mixture melts at around 950°C. Pure aluminium oxide melts at 2070°C. This temperature is too high for it to be used alone.
- Pour the liquid aluminium oxide and cryolite mixture into a huge electrolysis cell, as shown in the diagram below.
- Pass a current of 100 000 A through the liquid mixture.
  - Aluminium ions move to the negative electrode. Here, they gain electrons and form liquid aluminium metal.
  - Oxide ions move to the positive electrode. Here, they give up electrons to form oxygen gas. The oxygen reacts with the carbon electrode, making carbon dioxide gas.
- The half equation for the reaction at the negative electrode is $Al^{3+} + 3e^- \rightarrow Al$
- The half equation for the reaction at the positive electrode is $2O^{2-} - 4e^- \rightarrow O_2$

Key words

bauxite, cryolite

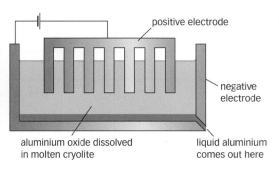

positive electrode

negative electrode

aluminium oxide dissolved in molten cryolite

liquid aluminium comes out here

◀ Aluminium electrolysis cell

Exam tip AQA

✔ Remember – the purpose of the cryolite is to reduce the temperature of the electrolysis cell. Aluminium forms at the negative electrode. Carbon dioxide forms at the positive electrode.

Questions

1 Name the substances formed at the positive and negative electrodes in the production of aluminium from aluminium oxide.

2 Name the ore from which most aluminium is extracted.

3 Describe what happens at the positive electrode during the electrolysis of aluminium oxide.

4 Write half equations for the reactions that occur at the positive and negative electrodes during the production of aluminium.

5 Suggest social, economic, and environmental benefits of recycling aluminium.

E

C

A*

Learning objectives

After studying this topic, you should be able to:

✔ predict the electrolysis products if there is a mixture of ions

✔ name and give uses of the products of the electrolysis of sodium chloride solution

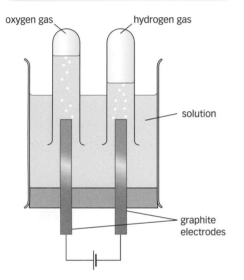

▲ Passing electricity through calcium sulfate solution produces oxygen and hydrogen gases

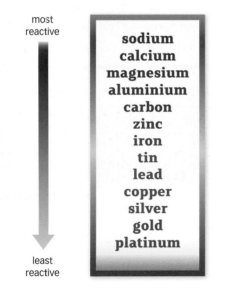

▲ Part of the reactivity series. The reactivity series lists metals in order of how vigorously they react with substances such as oxygen and water.

Products of electrolysis

If you electrolyse molten lead bromide, there is only one possible product at each electrode:

* lead at the negative electrode
* bromine at the positive electrode.

But for ionic compounds that are dissolved in water, predicting the electrolysis products is not quite so simple. Water takes part in electrolysis reactions, too.

Predicting products

Isabelle passed an electric current through some solutions, and identified the products at the electrodes.

The table summarises some of her results.

Solution	Product at negative electrode	Product at positive electrode
potassium iodide	hydrogen	iodine
magnesium bromide	hydrogen	bromine
copper nitrate	copper	oxygen
sodium carbonate	hydrogen	oxygen
calcium sulfate	hydrogen	oxygen

There is a pattern in the results.

* At the negative electrode:
  – The metal is produced if it is low in the reactivity series, like copper.
  – Hydrogen gas forms if the metal is above copper in the reactivity series. The hydrogen comes from the water.
* At the positive electrode:
  – If halide ions are present in solution, halogens will be produced.
  – If carbonate, sulfate, or nitrate ions are in the solution, oxygen is produced. The oxygen comes from the water.

A Predict the products of the electrolysis of sodium chloride solution.

B Predict the products of electrolysis of copper sulfate solution.

The electrolysis of sodium chloride solution

A whole industry has built up around the electrolysis of concentrated sodium chloride solution, or **brine**.

Passing electricity through brine produces these products:
- hydrogen gas at the negative electrode
- chlorine gas at the positive electrode.

A solution of sodium hydroxide also forms.

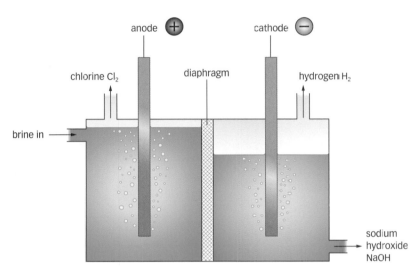

▲ The electrolysis of brine. The diaphragm keeps the chlorine away from the sodium hydroxide solution.

Using products from brine

The products of electrolysis of brine have many uses.
For example:
- Sodium hydroxide is used to make soap.
- Chlorine is used to make bleach and plastics, and to kill bacteria in drinking water and swimming pools.
- Hydrogen is used to make margarine and ammonia.

▲ Cells for the electrolysis of brine

Did you know...?
The European chlor-alkali industry produces around 10 million tonnes of chlorine a year. To do this, it uses 36 billion kilowatt hours of electricity – enough for about 7.5 million UK homes.

Key words
brine

Questions

1 Name the three products of the electrolysis of brine.
2 Give one use for each of the products of the electrolysis of brine.

E

3 Predict the products of the electrolysis of potassium bromide solution.
4 Predict the products of the electrolysis of melted potassium bromide.

C

5 Write a half equation for the reaction that happens at the positive electrode during the electrolysis of brine.
6 Write half equations for the reactions that happen at the electrodes during the electrolysis of copper bromide solution.

A*

Course catch-up

Revision checklist

- Rate of reaction is the change in amount of product (or reactant) per second.
- Reactions are faster at higher temperatures, at high concentrations and high pressure, and with powdered solids.
- Activation energy is the minimum amount of energy particles need to react.
- Catalysts increase the rate of a reaction but are unchanged at the end. They reduce the cost of industrial reactions.
- Some reactions (eg combustion) are exothermic (transfer energy to the surroundings).
- Some reactions (eg thermal decomposition) are endothermic (take in energy from the surroundings).
- Reversible reactions are endothermic in one direction and exothermic in the other.
- Acids have a pH of less than 7 and are neutralised by bases.
- Metal oxides or hydroxides are bases; ammonia is also a base.
- Alkalis are bases dissolved in water. They have a pH of greater than 7.
- Acids contain H^+ ions, alkalis contain OH^- ions. These react to form water.
- Salts form when acids react with bases, alkalis, or reactive metals.
- Soluble salts are prepared by neutralisation followed by crystallisation.
- Insoluble salts are made by precipitation reactions.
- Precipitation is used in water treatment.
- In electrolysis an electric current is passed through an ionic compound (molten or in solution). The compound breaks down into elements.
- Reduction happens at the negative electrode. Positive ions gain electrons.
- Oxidation happens at the positive electrode. Negative ions lose electrons.
- Metal objects are electroplated by making them a negative electrode and placing them in a solution of suitable ions.
- Aluminium is extracted by electrolysis of molten bauxite dissolved in cryolite.
- Electrolysis of brine produces chlorine, sodium hydroxide, and hydrogen.

combustion

mass of product formed per second

RATE OF CHEMICAL REACTIONS

reduce cost of industrial reactions

using a catalyst

alters rate but is unchanged at en

hydrogen, sodium hydroxide, and chlorine

sodium chloride

breaking down an ionic compound

ELECTROLYSIS

electroplating

of negative electrode dipped in silver ions, chronium ions etc

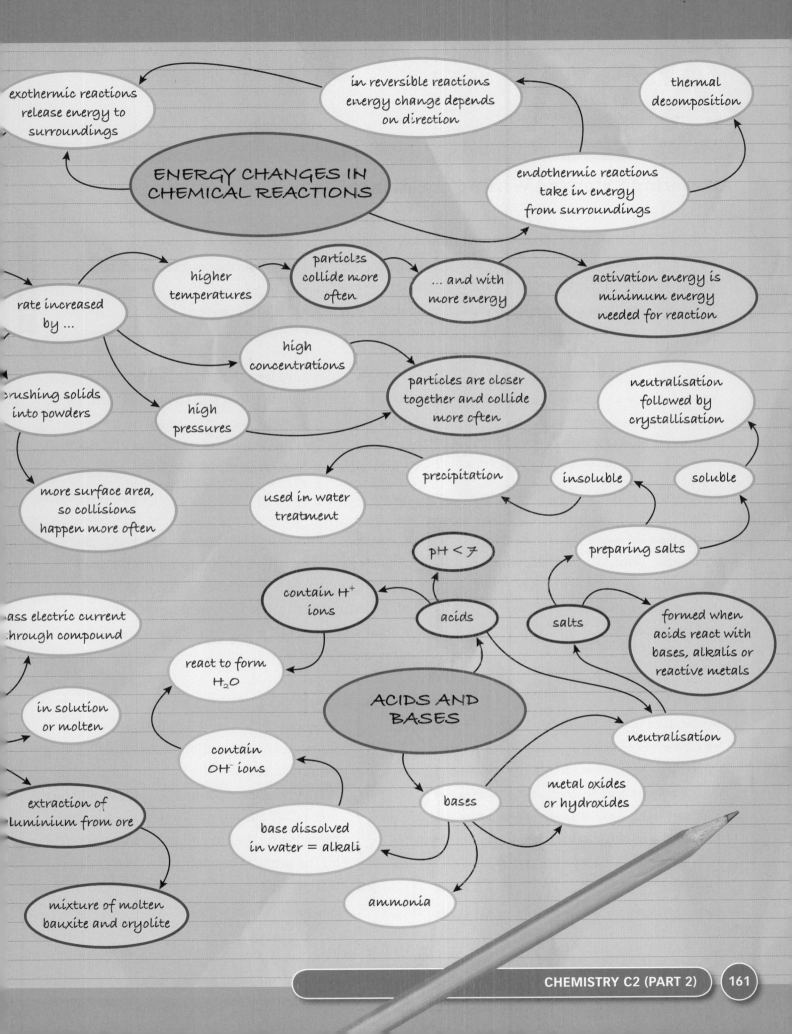

exothermic reactions release energy to surroundings

in reversible reactions energy change depends on direction

thermal decomposition

ENERGY CHANGES IN CHEMICAL REACTIONS

endothermic reactions take in energy from surroundings

rate increased by ...

higher temperatures

particles collide more often

... and with more energy

activation energy is minimum energy needed for reaction

high concentrations

crushing solids into powders

high pressures

particles are closer together and collide more often

neutralisation followed by crystallisation

more surface area, so collisions happen more often

used in water treatment

precipitation

insoluble

soluble

preparing salts

pH < 7

contain H⁺ ions

acids

salts

formed when acids react with bases, alkalis or reactive metals

ass electric current through compound

react to form H₂O

in solution or molten

ACIDS AND BASES

neutralisation

contain OH⁻ ions

metal oxides or hydroxides

extraction of luminium from ore

bases

base dissolved in water = alkali

ammonia

mixture of molten bauxite and cryolite

Answering Extended Writing questions

QUESTION

Acids and alkalis are two important types of chemical substance.

Describe the difference between these two substances and what happens when they react together. In your answer you should give an example of both an acid and an alkali.

The quality of written communication will be assessed in your answer to this question.

G–E

Acids, like hydrolic acid have a PH of below 7 and alkalis above 7. They cancel each other out when they react to make PH7.

Examiner: This candidate knows how to recognise acids and alkalis from pH values (but has shown pH incorrectly as 'PH'). However, the answer doesn't include the term 'neutralise'. The candidate has not named an alkali and isn't very close with the spelling of 'hydrochloric acid'.

D–C

Sulfuric acid has a pH of 1 and will newtralise sodium hydroxide which is an alkali and has a pH of over 7. When they react they make a salt, like sodium chloride.

Examiner: This candidate knows names of an acid and alkali and describes how they neutralise each other (though this is spelt wrongly). The answer should also mention that water is formed as well as a salt. It's a shame that the salt formed in the reaction discussed (sodium sulfate) is not named.

B–A*

Acids, like hydrochloric acid, have H+ ions and so have a pH of below 7. Alkalis are solubull bases, like sodium hydroxide and have OH ions and a pH of above 7. When they react the H+ and OH ions neutralise to make water and a salt (sodium chloride) is also made. The pH is now 7 because salt and water are neutral substances.

Examiner: This is a detailed and well-planned answer and includes an explanation about the ions are involved in neutralisation reactions. The candidate has missed off the charge on the OH⁻ ion, and there is one spelling mistake, but otherwise all the important scientific information is here.

Exam-style questions

1 Olivia heated 200 g of water using 1 g of two fuels, ethanol and hexane, in an experiment to help her decide which one is the best fuel.

Fuel	ethanol	hexane
Start temperature (°C)	21	21
End temperature (°C)	43	
Temperature rise (°C)		16

A03 **a** Complete the missing boxes.

A03 **b** Choose the correct description of the energy changes:
  i Both reactions give out energy.
  ii Hexane gives out energy and ethanol takes in energy.
  iii Hexane takes in energy and ethanol gives out energy.

A03 **c** Which fuel is the best?

A02 **d** Give one way in which Olivia made her experiment a fair test.

2 Jack investigated the factors affecting the reaction between magnesium and hydrochloric acid. He performed the reaction first with a low concentration of hydrochloric acid, and then with a high concentration.

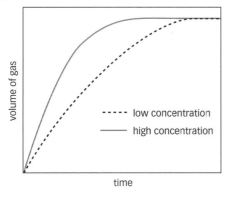

A03 **a** Which reaction has the fastest rate?

A03 **b** What can Jack conclude about the effect of increasing concentration on rate of reaction?

A02 **c** In another experiment, Jack found that increasing the temperature made the reaction faster. Use ideas about particles to explain why.

3 Hydrochloric acid can be neutralised by a solution of calcium hydroxide. A salt and water are formed.

A02 **a** Name the salt that is formed.

A01 **b** What name is given to soluble bases such as calcium hydroxide?

A02 **c** Write a balanced symbol equation to show how the hydroxide ions in calcium hydroxide are neutralised by ions from the hydrochloric acid.

Extended Writing

4 Catalysts are often used in industrial reactions. What are catalysts and why are they important in industry? Give an example of their use in industry.
A01

5 Shafiq wants to make a sample of solid magnesium sulfate, a soluble salt, from solid magnesium oxide and sulfuric acid. Write a set of instructions that he can follow to make this salt.
A02

6 Electrolysis of brine (sodium chloride) is used to make hydrogen, sodium hydroxide, and chlorine. Describe what happens in this process and why it is important.
A01

G–E

D–C

B–A*

A01 Recall the science
A02 Apply your knowledge
A03 Evaluate and analyse the evidence

CHEMISTRY C2 (PART 2)   163

P2 Part 1

Forces and motion

Why study this unit?

You can use physics to describe the motion of objects, and you can also use it to predict what will happen to an object in many different conditions. When objects move, energy transfers take place, for example from gravitational potential energy to kinetic energy when you drop an object and it falls to the floor. When engineers are designing cars, they need to be able to predict what will happen to the car and its occupants in order to minimise possible injuries if the car is in a crash.

In this unit you will look at how the acceleration of an object is linked to the force acting on it, and how this can change its motion. You will learn how the motion of the object can be represented in graphs, and how the object's motion is affected by air resistance. You will also learn about the distance it takes to stop a car, and how this distance is affected by different conditions and by the state of the driver. You will learn about momentum, and how this influences the design of car safety features.

You should remember

1 When a force acts on an object it can cause it to move.

2 Friction is a force that tries to stop things from moving.

3 When an object is moving, air resistance tries to slow it down.

4 Energy cannot be created or destroyed.

5 Energy can exist in different forms, such as kinetic energy and gravitational potential energy.

The world's fastest rollercoaster is the Ring Racer at the Nürburgring race track in Germany. It has a top speed of 217 km/h and accelerates from 0 to 217 km/h in 2.5 seconds. It has been designed to simulate the speed of a Formula 1 racing car. It does not have any loops or banked turns and has been designed simply to travel at high speeds. Engineers will have considered the forces acting on the rollercoaster and riders to give the acceleration they need to reach the top speed. They will also have considered how to keep the riders safe by including many safety features.

The Ring Racer operated briefly in 2009 until it was damaged by an explosion in the control system. It is now scheduled to open to the public in 2011. There is an even faster roller coaster under construction in Dubai – its top speed will be 240 km/h.

Learning objectives

After studying this topic, you should be able to:

- ✔ describe how when two objects interact, the forces they exert on each other are equal and opposite
- ✔ explain the meaning of the term resultant force
- ✔ understand effects of forces on an object

▲ On take-off the Space Shuttle's engines provide a thrust that pushes against the pull of gravity

▲ Pushing against a wall creates a pair of forces that are equal and opposite

What are forces?

Forces are all the different pushes or pulls which are around us all of the time. Everyday forces include air resistance, gravity and **friction**. All forces are measured in **newtons**. A teenager might be able to lift a weight of 800 N and a large jet engine might produce a thrust of 600 000 N.

> **A** Give three examples of forces.

Pairs of forces between two objects

Whenever two objects interact, they exert forces on each other. They push onto each other and this produces a **pair of forces**. These pairs of forces are always **equal** in size and **opposite** in direction. You can feel this if you push down on the desk. It pushes back up at you with an equal and opposite force.

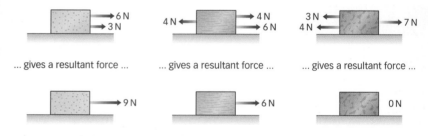

▲ Calculating the resultant force acting on an object

Resultant force

There is often more than one force acting on an object. When a number of different forces act at a point on an object, they add up to a **resultant force**. This is the single force that has the same effect as all the original forces combined.

To work out the resultant force in a direction, you add up the forces acting in a straight line in that direction.

Any forces acting in the opposite direction need to be subtracted from the total, and so a number of forces might cancel out. This might give a resultant force of zero.

> **B** Explain what is meant by resultant force.

What do forces do?

A resultant force is not needed to keep an object moving. When a car travels at a steady speed along a straight motorway, the force from the engine is equal to the resistive forces (air resistance and friction). There is no resultant force, it is zero, yet the car continues to move.

A resultant force *changes* the way something moves. If an object is not moving (stationary), then a resultant force makes it begin to move, and if it is already moving, a resultant force changes the motion of the object. If the driver takes their foot off the accelerator then there is a resultant force (backwards) and the car slows down. If the driver pushes their foot down then the force from the engine is greater than the resistive forces, there is a resultant force forwards and the car gets faster.

A resultant force makes an object **accelerate**, it will make it speed up, slow down or change direction.

	Zero resultant force	Resultant force
Stationary object	Remains stationary	Starts moving (accelerates)
Moving object	Continues moving at a steady speed in a straight line	Accelerates in the direction of the resultant force (this might mean slowing down if the force is in the opposite direction to the motion of the object)

Questions

1 List two things forces can do to objects.

2 Which two words describe the pair of forces produced when two objects interact? Explain what they mean.

3 Describe what happens to an object if there is resultant force acting on it.

4 Draw diagrams to show the forces acting on:
   (a) A cyclist getting faster
   (b) A cyclist travelling at a steady speed
   (c) A cyclist getting slower.

E

↓
C

↓
A*

Key words

force, friction, newton, pair of forces, equal and opposite, resultant force, accelerate

Did you know...?

On Earth when you hit a tennis ball both gravity and air resistance act on the ball. This causes the ball to slow down and change direction (bend towards the Earth). A tennis ball that was hit in deep space would continue in a straight line at a steady speed. There are no forces to slow it down or change its direction.

▲ Skilled tennis players have mastered the forces acting on the ball

Exam tip

✔ Remember, forces don't make objects move. If there is a resultant force on an object, the object will change the way it is moving (accelerate).

Learning objectives

After studying this topic, you should be able to:

- ✔ understand how forces change the motion of objects
- ✔ use the equation resultant force = mass × acceleration

Key words

acceleration

▲ At the sound of the starter's gun, this sprinter's legs provide a force that accelerates him out of the blocks

A What is acceleration?

B Describe the relationship of acceleration and mass.

Forces and acceleration

You have already learnt that a resultant force will make an object accelerate. Forces such as friction or the thrust from an engine change the way an object moves. We will look at **acceleration** in more detail later, but you can think of acceleration as a change in speed or a change in direction.

If you double the resultant force acting on an object, its acceleration will double (as long as you keep the mass of the object the same). Applying a resultant force of 12 N might make an object accelerate at 3 m/s². Increasing this force to 24 N will make the same object accelerate at 6 m/s².

If you apply the same force to different objects, they may accelerate at different rates. The greater the mass of the object, the lower its acceleration. Imagine the engines in the vehicles shown below can all provide the same force. The bike will accelerate at the greatest rate. This is because the bike has the lowest mass. The lorry has the greatest mass and so it will accelerate at the lowest rate.

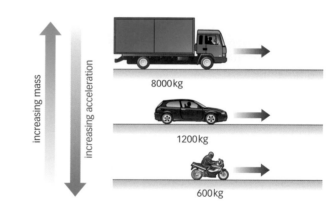

8000 kg

1200 kg

600 kg

▲ The greater the mass of an object, the lower its acceleration for a particular force

Force = mass × acceleration

The resultant force acting on an object, its mass, and its acceleration are related in this equation:

resultant force = mass × acceleration
(newtons, N) (kilograms, kg) (metres per second², m/s²)

If the resultant force is called F, the mass m, and the acceleration a, then:

$$a = \frac{F}{m} \quad \text{or} \quad F = m \times a$$

Worked example 1

A sprinter accelerates out of the blocks at 3.0 m/s². She has a mass of 70 kg. What is the resultant force from her legs?

resultant force = mass × acceleration

$$F = m \times a$$

m = 70 kg and a = 3 m/s²

resultant force = 70 kg × 3 m/s²

$$= 210 \text{ N}$$

Worked example 2

A passenger jet has mass of 320 000 kg and it has 200 000 N of thrust from each of its four engines. Calculate its acceleration.

resultant force = mass × acceleration, so

$$\text{acceleration} = \frac{\text{resultant force}}{\text{mass}} \quad \text{or} \quad a = \frac{F}{m}$$

mass = 320 000 kg

total resultant force = 4 × 200 000 N = 800 000 N

$$\text{acceleration} = \frac{800\,000 \text{ N}}{320\,000 \text{ kg}}$$

$$= 2.5 \text{ m/s}^2$$

Questions

1. State the equation that links resultant force, mass and acceleration, including units for each term. ↓ E

2. A charging rhino has a mass of 1400 kg and accelerates at 1.5 m/s². Calculate the resultant force providing the acceleration.

3. Sketch a graph of resultant force against the acceleration of an object. ↓ C

4. A football with a mass of 400 g is kicked with a force of 700 N. Calculate the ball's acceleration.

5. A small rocket has a weight of 200 N and a mass of 20 kg. When launched it accelerates at 5 m/s². Find: ↓ A*

   (a) the resultant force

   (b) the thrust from the engine.

Did you know...?

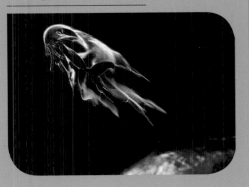

▲ The mass of the flea is so small that the force from its legs provides a gigantic acceleration

Acceleration can be measured in 'G'. An acceleration of 2 G would be twice the acceleration due to gravity. Racing drivers often experience large accelerations of 4 G or 5 G when they take tight bends. When a flea jumps, it accelerates at over 1200 m/s². That's 120 G. Even the most experienced fighter pilots would pass out at around 8 G. The flea's legs produce only a small force, but the acceleration is huge because the flea has such a tiny mass.

Exam tip AQA

✓ When using the equation $F = m \times a$, you must remember to use the resultant force. You may need to calculate this first.

Learning objectives

After studying this topic, you should be able to:

✔ calculate the speed of an object

✔ explain the difference between speed and velocity

✔ describe how cameras are used to measure speed

Speed

Speed is a measure of how fast someone or something is moving. It is the distance moved in a certain time. It is calculated using the equation:

$$\text{average speed (metres/second, m/s)} = \frac{\text{distance (metres, m)}}{\text{time (seconds, s)}}$$

For example, in a sprint race the athletes run a measured distance, and the time they take to run the distance is also measured. So you can work out their speed.

◀ We can work out the speeds of these athletes because we know how long it took them to run a certain distance

Worked example

Usain Bolt ran 100 metres in 9.58 seconds. On average, how fast did he run?

$$\text{average speed} = \frac{\text{distance}}{\text{time}}$$

distance = 100 m and time = 9.58 s, so

$$\text{average speed} = \frac{100 \text{ m}}{9.58 \text{ s}}$$

$$= 10.4 \text{ m/s}$$

Both the distance and the time need to be measured accurately to get an accurate measure of speed. You can measure distances using a surveyor's tape or a trundle wheel. The length of the circumference of the wheel is known, and the number of times the wheel rotates is counted. The time taken to move the measured distance can be measured using a stopwatch.

Speeds can also be measured in other units as well as in seconds. Speeds of cars and other vehicles are often measured in miles per hour (mph) or kilometres per hour (km/h).

◀ This man is using a trundle wheel to measure distances

Speed cameras

Speed cameras are used to measure the speeds of vehicles that are travelling faster than the speed limit.

Some speed cameras are used together with lines painted on the road, as shown in the picture on the next page. As the car passes over the lines, the camera takes two pictures 0.2 seconds apart. The distance travelled by the vehicle in that time is found by looking at the two photos. The speed is then calculated using the equation, speed = distance/time.

A A car travels 1000 metres in 40 seconds. What is the speed of the car?

B Usain Bolt ran 200 metres in 19.19 seconds. How fast did he run?

▲ A speed camera. The lines painted on the road are usually 1 m apart. The speed of the car is worked out by measuring the distance the car travels in a certain time.

A pair of cameras can also be used to work out the average speed of a car. The time when the car passes each camera is recorded. The distance between the cameras is known, so the car's speed can be worked out.

Velocity

Speed tells you how fast something is moving, but it does not tell you what direction it is moving in. **Velocity** tells you the direction an object is travelling in as well as its speed. For example, you might say that a car was moving north at 30 km/h.

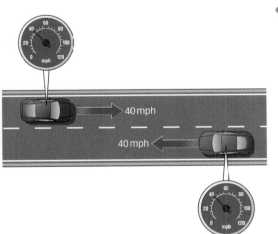

◀ These cars may be travelling at the same speed, but they have different velocities

The two cars in the picture are both travelling at 40 mph; their speeds are the same. But they are moving in different directions, so their velocities are different.

We might say that the red car has a velocity of +40 mph. The blue car is travelling at 40 mph in exactly the opposite direction. We would say that it has a velocity of –40 mph.

Exam tip

✔ When calculating speed, take care with the units.

✔ Remember that speed and velocity are not the same thing. Velocity gives the direction of travel as well as the size of the speed.

C A car travels 4 metres in 0.2 seconds. How fast is the car travelling?

Questions

1 A cyclist covers a distance of 100 metres in 20 seconds. What is his speed in metres/second? ↓ E

2 A car travels 240 km in 3 hours. What is the speed of the car in km/h?

3 One car is travelling north at 70 mph. Another car is travelling eastwards at 70 mph. Do the cars have the same velocity? Explain your answer. ↓ C

4 Explain how a speed camera and painted lines on the road are used to find the speed of a car.

5 The car in Question C is in an area where the speed limit is 60 km/h. Is it breaking the speed limit? ↓ A*

Learning objectives

After studying this topic, you should be able to:

✔ draw distance–time graphs

✔ understand that the gradient of a distance–time graph represents speed

✔ calculate the speed of an object from a distance–time graph

Recording distance and time

You can record the distances that an object travels and the time taken to travel those distances. The table in the margin shows the distance a car has travelled along a motorway. The distance and time are measured from where and when the car started.

You can plot this data on a **distance–time graph**. Time is usually plotted on the *x*-axis and distance on the *y*-axis.

Gradient

You can tell how fast something is moving by looking at the slope of the line. If the car is moving faster, it goes a greater distance in every 50 seconds and the slope of the line is steeper. If the car is slower, it moves a smaller distance every 50 seconds and the slope is less steep.

We call the slope the **gradient** of the graph. The gradient of a distance–time graph represents speed.

If a distance–time graph has a straight slope, this tells you that the object is moving at a constant speed. Where the line in a distance–time graph is horizontal, the object has not moved any distance – it is **stationary**.

Time (s)	Distance (m)
0	0
50	1500
100	3000
150	4500
200	6000
250	7500
300	9000

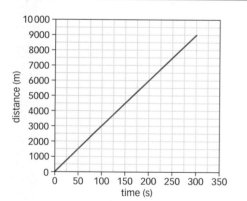

▲ Distance–time graph for the car on the motorway, using the data in the table

Exam tip **AQA**

✔ When calculating speed from a distance–time graph, remember to find the change (difference) in distance and the change in time.

A The table shows the time and distance measurements for a contestant in an athletics race. Draw a distance–time graph for the data shown in the table.

Time taken (s)	0	20	40	60	80	100	120
Distance run (m)	0	130	260	390	520	650	780

B Look at this distance–time graph for two runners in a race. Which runner is faster?

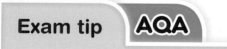

Calculating speed from a distance–time graph

You can calculate the speed of an object by finding the gradient of a distance–time graph. You can work out the gradient using the equation:

$$\text{gradient (speed)} = \frac{\text{difference in distance}}{\text{difference in time}}$$

Worked example

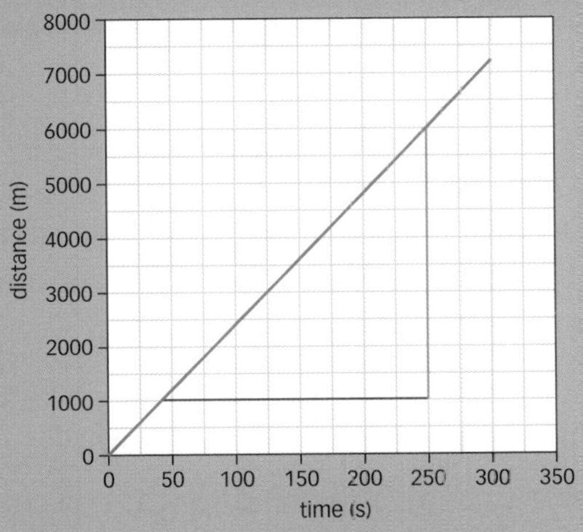

What is the speed of the car shown on the distance–time graph (blue line)?

Draw a right-angled triangle under the graph line.

difference in distance (green line): 6000 m – 1000 m = 5000 m

difference in time (red line): 250 s – 40 s = 210 s

$$\text{gradient (speed)} = \frac{\text{difference in distance}}{\text{difference in time}}$$

$$= \frac{5000 \text{ m}}{210 \text{ s}}$$

$$= 23.8 \text{ m/s}$$

Did you know...?

All new lorries are fitted with a tachograph. This records the speed at which a lorry has travelled over time. They can be used to check that lorry drivers have taken the breaks that are required by law and to make sure that they have kept to the speed limit.

Key words

distance–time graph, gradient, stationary

Questions

1 What does the gradient of a distance–time graph tell you? ↓E

2 The line of a distance–time graph slopes steeply up from left to right. What does this tell you about the motion of the object?

3 What does a horizontal line on a distance–time graph tell you?

4 A cyclist sets out on a straight road. After 50 seconds she has travelled 200 m. She stops for 100 seconds to adjust her bike. She then travels 1000 m in 200 seconds. Draw the distance–time graph for her journey. ↓C

5 Find the speeds of the runners as shown in the graph in Question B. ↓A*

Learning objectives

After studying this topic, you should be able to:

✔ calculate the acceleration of an object

▲ A cheetah can speed up from rest to 20 m/s in less than 2 seconds. That's a greater acceleration than most cars are capable of.

▲ This car is accelerating because it is going round a bend

Exam tip **AQA**

✔ Remember that slowing down is also acceleration in scientific language.

✔ When you are working out acceleration, don't forget to calculate the change in velocity – don't just use the final velocity.

Speeding up and slowing down

A moving object might speed up or slow down. This change in speed is called acceleration. The change can be negative as well as positive. When something is slowing down, it will have a negative acceleration. In everyday language negative acceleration is called **deceleration**.

> **A** What is deceleration?

Acceleration also has a direction, like velocity. For example, when a car is pulling away from traffic lights, the acceleration is in the same direction that the car is moving in, and it is positive. When the car slows down at another set of traffic lights, the acceleration is in the opposite direction to the car's motion and it is negative.

If a car has a negative acceleration this can mean that the car is slowing down or even moving backwards.

When the velocity of an object changes, it is accelerating. A change in velocity can also mean a change in direction. Even if the speed stays the same, but the direction changes, the object is being accelerated.

> **B** Why can we say that the car in the photo is accelerating?

Calculating acceleration

Acceleration is the rate at which velocity changes. It depends on how much the velocity changes and the time taken for the change of velocity.

$$\text{acceleration (metres per second squared, m/s}^2) = \frac{\text{change in velocity (metres per second, m/s)}}{\text{time taken for change (seconds, s)}}$$

The acceleration is called a, the initial velocity is called u, the final velocity is called v, and the time taken for the change is called t.

Change in velocity is final velocity – initial velocity = $v - u$, so the acceleration is:

$$a = \frac{v - u}{t}$$

Worked example 1

The velocity of a train increases from 15 m/s to 35 m/s in 10 seconds. What is the acceleration of the train?

$$\text{acceleration} = \frac{\text{change in velocity}}{\text{time taken for change}} \quad \text{or } a = \frac{v-u}{t}$$

initial velocity, u = 15 m/s

final velocity, v = 35 m/s

time taken for change, s = 10 seconds

$$\text{acceleration} = \frac{35 \text{ m/s} - 15 \text{ m/s}}{10 \text{ s}} = \frac{20 \text{ m/s}}{10 \text{ s}} = 2 \text{ m/s}^2$$

C A cheetah accelerates from 3 m/s to 18 m/s in 3 seconds. What is its acceleration?

Acceleration is negative when an object is slowing down or if it starts to move backwards.

Worked example 2

A car slows down and starts moving backwards. Its velocity changes from 20 m/s to –5 m/s in 10 seconds. What is the acceleration of the car?

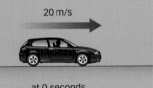 

20 m/s   –5 m/s

at 0 seconds   at 10 seconds later

Draw a diagram to help understand what is happening.

$$\text{acceleration} = \frac{\text{change in velocity}}{\text{time taken for change}} \quad \text{or } \quad a = \frac{v-u}{t}$$

initial velocity, u = 20 m/s

final velocity, v = –5 m/s

time taken for change, s = 10 seconds

$$a = \frac{v-u}{t} = \frac{-5 \text{ m/s} - 20 \text{ m/s}}{10 \text{ s}} = \frac{-25 \text{ m/s}}{10 \text{ s}} = -2.5 \text{ m/s}^2$$

The value for the acceleration is negative because the car was slowing down and then moving in the opposite direction.

Key words

deceleration

Did you know...?

It is the effects of acceleration that provide the thrills on a roller coaster.

▲ This roller coaster will accelerate as it moves down the track

Questions

1 What is acceleration?

2 What does it mean when an object has negative acceleration?

3 The speed of a train increases from 5 m/s to 55 m/s in 20 seconds. What is the acceleration of the train?

4 When an aircraft lands, its speed is 65 m/s. The speed decreases to 10 m/s in 11 seconds. What is the acceleration of the aircraft?

5 A force of –5 N acts on an object of mass 2 kg moving at 10 m/s. After how many seconds will the object start to move backwards?

Learning objectives

After studying this topic, you should be able to:

✔ draw a velocity–time graph

✔ explain that the gradient of a velocity–time graph represents acceleration

✔ calculate the acceleration by using the gradient of a velocity–time graph

✔ use a velocity–time graph to calculate the distance travelled

Key words

velocity–time graph

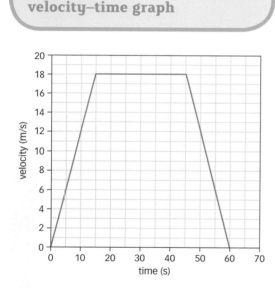

▲ Velocity–time graph for a train travelling in a straight line between two stations

A What does a velocity–time graph show?

B What does a horizontal line on a velocity–time graph mean?

Using velocity–time graphs

In the same way that you can record an object's distance at different times, you can also record an object's velocity at different times. The graph on the left tells you how fast a train is moving, whether it is speeding up or slowing down, and whether it is moving forwards or backwards.

A **velocity–time graph** usually has time on the *x*-axis and velocity on the *y*-axis:

- If an object is moving at a steady (constant) velocity, the line on the graph is horizontal; the velocity is not changing.
- If the object is going steadily faster, the velocity is steadily increasing, and the graph shows a straight line sloping upwards.
- If the object is slowing down steadily, the velocity is steadily decreasing, and the graph shows a straight line sloping downwards.
- If the object moves backwards, the velocity is negative, and so those graph points will be plotted below the *x*-axis.

The slope or gradient of a velocity–time graph is the change in velocity divided by the change in time. This is acceleration.

The steeper the slope, the greater the acceleration.

When the graph slopes downwards to the right, because the object is slowing down, the acceleration is negative. We also say that the gradient is negative.

Calculating acceleration

You can calculate the acceleration by working out the gradient of a velocity–time graph. The gradient is given by the equation:

$$\text{gradient (acceleration)} = \frac{\text{change in velocity}}{\text{time taken for change}}$$

This is the same as the equation for acceleration that you met on spread P2.5:

$$a = \frac{v - u}{t}$$

Distance travelled

The area under a velocity–time graph shows the distance travelled.

Worked example 1

These are velocity–time graphs for two cars. Calculate the acceleration of car A.

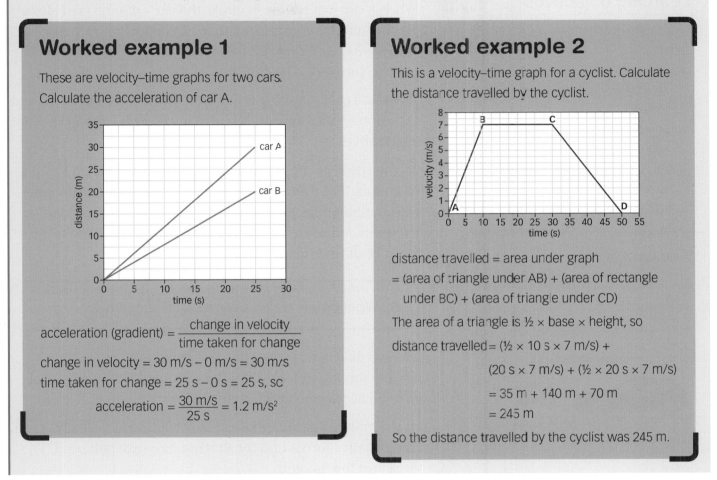

$$\text{acceleration (gradient)} = \frac{\text{change in velocity}}{\text{time taken for change}}$$

change in velocity = 30 m/s – 0 m/s = 30 m/s

time taken for change = 25 s – 0 s = 25 s, sc

$$\text{acceleration} = \frac{30 \text{ m/s}}{25 \text{ s}} = 1.2 \text{ m/s}^2$$

Worked example 2

This is a velocity–time graph for a cyclist. Calculate the distance travelled by the cyclist.

distance travelled = area under graph

= (area of triangle under AB) + (area of rectangle under BC) + (area of triangle under CD)

The area of a triangle is ½ × base × height, so

distance travelled = (½ × 10 s × 7 m/s) +

(20 s × 7 m/s) + (½ × 20 s × 7 m/s)

= 35 m + 140 m + 70 m

= 245 m

So the distance travelled by the cyclist was 245 m.

Questions

1 What does the gradient of a velocity–time graph show?

2 Look at the velocity–time graphs for the two cars in the worked example. Does car A or car B have the greater acceleration? Explain your answer.

3 Draw a velocity–time graph for:

(a) a person walking at a constant velocity of 1 m/s for 10 seconds

(b) an aircraft accelerating from 0 m/s to 60 m/s over 20 seconds.

4 Look at the velocity–time graph for the train on the previous page. Calculate the acceleration of the train as shown in each part of the graph.

5 Look at the velocity–time graph for the car that stops and then reverses. Calculate the distance travelled by the car between 0 s and 4 s.

E

C

A*

Learning objectives

After studying this topic, you should be able to:

✔ understand how the forces acting on a vehicle are balanced when it travels at a steady velocity

✔ explain that the braking force needed to stop a vehicle in a certain distance increases as the speed increases

Key words

air resistance, resistive force, braking force

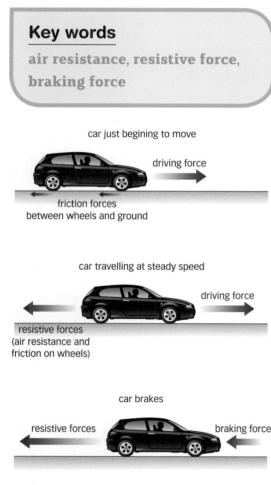

car just begining to move

driving force

friction forces between wheels and ground

car travelling at steady speed

driving force

resistive forces (air resistance and friction on wheels)

car brakes

resistive forces

braking force

▲ Forces acting on a moving car

Forces on a car

When a car just begins to move, the car's engine provides a driving force that is greater than the opposing friction forces on the wheels. The resultant force on the car is not zero. The car accelerates in the direction of the resultant force.

When the car is moving it has to push air out of the way, and the air exerts a force on the car. This force is called **air resistance**.

As the speed of the car increases, the air resistance increases. The driving force of the car must still overcome the friction forces between the wheels and the road, and frictional forces in the engine, but the air resistance is the main **resistive force**.

> **A** What forces act on a car when it is moving?

The top diagram shows a car just starting to move. The driving force is greater than the frictional force and so the car will accelerate to the right.

In the middle diagram, the car is travelling at a steady speed. The resistive forces and the driving force are balanced. The resultant force is zero.

In the bottom diagram, the resultant force is backwards. The car is slowing down. The resultant backwards force comes from removing the driving force and applying a **braking force**.

A car engine has a maximum force that it can provide. As the speed of a car increases, the resistive forces increase. When the car reaches a certain speed, the resistive forces will be equal to the driving force and the car will not be able to go any faster.

> **B** What happens to the resistive forces as the speed increases?
>
> **C** What is the direction of the resultant force when the car slows down?

Stopping a car

When a car brakes, a force is being applied to slow the car down. If the car has to stop in a certain distance, the force needs to be greater if the car is travelling faster. If the braking force is always the same (remains constant), then the distance needed to stop the vehicle increases as the speed of the vehicle increases.

▲ The shape of this train has been designed to reduce air resistance

The acceleration and distance needed to stop at different speeds. The acceleration is in the opposite direction to the motion of the car, so it is negative (in other words, a deceleration).

Questions

1 What is the resultant force on a car that is moving at a constant speed? **↓ E**

2 What is the braking force on a car that is moving at a constant speed?

3 A car must stop in a certain distance. Explain the link between the speed of the car and the braking force needed. **↓ C**

4 The same constant braking force is used to stop a car, whatever speed it is travelling at to start with. Explain what happens to the car's stopping distance for higher speeds.

5 Explain how the shape of the train shown in the photo enables it to travel faster. **↓ A***

Did you know...?

As faster vehicles are designed, more streamlining is needed to reduce air resistance. The train in the photo was built in Japan in 2009 and is designed to travel at 320 km/h.

Exam tip

✔ Remember that the acceleration will always be in the same direction as the resultant force.

Learning objectives

After studying this topic, you should be able to:

✔ explain what the stopping distance of a vehicle is

✔ state what affects the stopping distance

✔ describe the energy transfers that take place when a car is braked

---

A What is stopping distance?

Exam tip

✔ You don't need to remember actual stopping distances, but you do need to remember what can affect them, such as reaction time, condition of vehicle and road conditions.

▲ Using a mobile phone in this way while driving can distract you. It is also illegal.

Key words

braking distance, reaction time, thinking distance, stopping distance

Stopping distances

The total distance needed to stop a car is not just the distance the car travels after the brakes have been applied, called the **braking distance**. There is also the time needed for the driver to react to seeing something. For example, the driver sees a red light and needs to move their foot onto the brake pedal. This is called the **reaction time**. During the reaction time, the car will have travelled a certain distance called the **thinking distance**.

So the total **stopping distance** is made up of the thinking distance and the braking distance.

The Highway Code gives stopping distances under normal conditions, as shown in the diagram. The calculations assume that the acceleration is –6 m/s². (The minus sign means a negative acceleration, or deceleration. It is in the opposite direction to the direction of motion.)

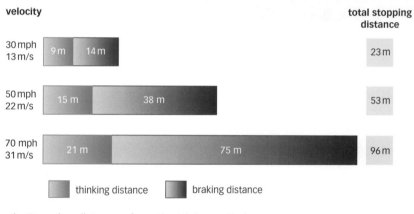

velocity			total stopping distance
30 mph 13 m/s	9 m	14 m	23 m
50 mph 22 m/s	15 m	38 m	53 m
70 mph 31 m/s	21 m	75 m	96 m

☐ thinking distance   ☐ braking distance

▲ Stopping distances from the Highway Code

Thinking distance

There are many factors that can affect your reaction time and hence the thinking distance. When you are tired, you react more slowly. If you have used drugs such as alcohol or illegal drugs, your reactions are slower. Some drugs that are available over the counter or prescription drugs can also increase the time it takes you to react.

Distractions, such as listening to music, using a mobile phone or a satellite navigation system, or even talking to passengers, can increase reaction time. Also, people's reactions become slower as they get older.

Braking distance

The braking distance does not only depend on the speed of the car. When you press the brake pedal, the brakes are applied to the wheel. If the brakes are worn, this can reduce the force that they can apply. If too much force is applied, the wheels can lock and the car skids.

Road conditions can also affect the braking distance. If the road surface is icy, there is less friction between the tyres and the road and the tyres may slip. The braking distance will increase. On wet road surfaces the friction between the tyres and the road is also reduced.

▲ The back marks on the runway are from the tyres of planes

▲ Stopping distances are longer in conditions like this

> **B** What happens when you press the brake pedal?
>
> **C** Why do brakes need to be in good condition?

Energy transfers

When you apply the brakes on a car, the frictional force is increased. Kinetic energy is transferred into heat energy in the brakes.

If the wheels of a vehicle lock when the vehicle is braking, the tyres will skid along the road. There is then much more friction between the tyre and the road surface, and much more kinetic energy is transferred into heat energy.

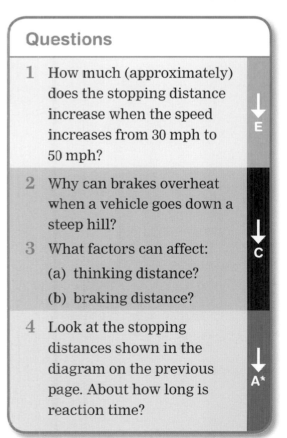

Did you know...?

There are always skid marks where planes land at the end of airport runways. The plane's wheels are not moving at the moment that they touch the runway. When they touch the runway, they skid briefly. A small amount of smoke is produced – kinetic energy is transferred to heat energy, which burns a small amount of rubber.

Questions

1 How much (approximately) does the stopping distance increase when the speed increases from 30 mph to 50 mph? ↓E

2 Why can brakes overheat when a vehicle goes down a steep hill?

3 What factors can affect: ↓C
  (a) thinking distance?
  (b) braking distance?

4 Look at the stopping distances shown in the diagram on the previous page. About how long is reaction time? ↓A*

Learning objectives

After studying this topic, you should be able to:

✔ calculate the weight of an object

✔ describe the motion of an object falling under gravity

Key words

weight, mass, gravitational field strength, fluid

Worked example

A large bag of rice has a mass of 2 kg. What is its weight?

weight = mass × gravitational field strength

$\qquad$ = 2 kg × 10 N/kg

$\qquad$ = 20 N

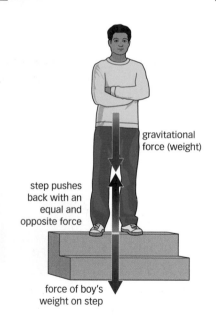

gravitational force (weight)

step pushes back with an equal and opposite force

force of boy's weight on step

▲ The resultant force on the boy is zero

Weight

When you drop an object, it falls directly towards the ground. It falls because of the gravitational attraction between the object and the earth. It is pulled by the force of the Earth's gravitational field. This force is called **weight**.

The weight of an object depends on two things:

• its **mass** (the amount of matter) in kilograms (kg)
• the **gravitational field strength** in newtons/kilogram (N/kg).

The equation linking weight, mass, and gravitational field strength is:

$$W \qquad = \qquad m \qquad \times \qquad g$$

| weight of object (newtons, N) | = | mass of object (kilograms, kg) | × | gravitational field strength (newtons per kilogram, N/kg) |

The Earth's gravitational field strength is about 10 N/kg.

> **A** Tom has a mass of 50 kg. What is his weight?
>
> **B** A car has a mass of 1450 kg. What is its weight?

Falling under gravity

In the picture, the boy and the step are interacting objects that produce an equal and opposite pair of forces on each other.

To decide how the boy will move, we look only at the forces acting on the boy. There are two of them: the gravitational attraction downwards to the Earth, and the upward force from the step. These are the same size, the resultant force is zero and the boy is not accelerating. He is also not moving.

> **C** What two forces are acting on the boy in the diagram?

If the boy moves forward off the step, there is no upward force acting on him, and his weight will cause him to accelerate downwards. The ball in the diagram on the next page does not have anything to stop it from falling towards the Earth. The force of gravity makes the ball accelerate downwards.

As the ball falls, its velocity continues to increase.

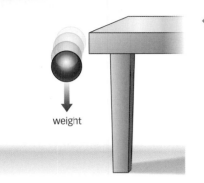

The resultant force on this ball is downward

weight

Did you know...?

Gravitational attraction acts both ways. So when the leaf and the ball were falling downwards, they would actually be pulling the Earth a little bit upwards! But the effect would be much too small for us to notice.

Air resistance

When something is falling through the air, as its velocity increases, the size of the force in the opposite direction due to air resistance increases. The faster an object moves through air, or any other **fluid**, the greater the resistive force that acts on the object in the opposite direction.

The diagram on the right shows what happens to a falling leaf. When the leaf first begins to fall, the upward force due to air resistance is low. As the speed of the leaf increases, so does the air resistance. The downward force from the weight of the leaf stays the same.

The size of the air resistance also depends on the shape and the surface area of the object. The air resistance force is much greater for the leaf than for the ball shown earlier.

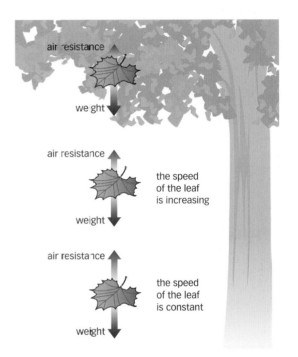

air resistance

weight

air resistance — the speed of the leaf is increasing

weight

air resistance — the speed of the leaf is constant

weight

▲ As the leaf falls more quickly (accelerates due to gravitational attraction), the air resistance gets bigger

> **D** What happens to the force of air resistance when speed increases as an object falls under gravity?

Questions

1 What is the weight of an object?

2 You drop a pen and it falls to the floor. Describe the motion of the pen.

3 Calculate the weight of each of the following:
   (a) a table with mass 25 kg
   (b) an elephant with mass 692 kg.

4 Alex says that his weight is 65 kg. Explain why he is wrong.

5 An object has a weight of 12 N. What is its mass?

Exam tip   AQA

✓ In everyday language, the term 'weight' is used to mean 'mass' – when you talk about someone's weight, you usually mean their mass. Make sure you know the difference between the two and use them correctly.

Learning objectives

After studying this topic, you should be able to:

✔ understand how a falling object reaches a terminal velocity

✔ draw and interpret velocity–time graphs for objects that reach terminal velocity

Key words

terminal velocity

A What forces are acting on the skydiver in the pictures?

B What is the terminal velocity of an falling object?

C What can you say about the forces acting on anything moving at a terminal velocity?

✔ When the parachute opens, the skydiver's velocity decreases because there is an acceleration that acts upwards. The skydiver does not start to move upwards – they continue to fall, but at a lower velocity.

Terminal velocity

You already know that when something falls downwards due to gravity, it accelerates and its velocity increases. As its velocity increases, the upward force of air resistance increases, as shown for the skydiver in the diagram.

▲ Skydiver reaching a terminal velocity: resultant forces and velocity–time graph

As the skydiver speeds up, the air resistance increases. The skydiver's weight stays constant.

A There is still a resultant force downwards, so there is still an acceleration downwards and the velocity is still increasing.

B Because the velocity is still increasing, the air resistance gets bigger. The resultant force downwards gets smaller. The acceleration downwards also decreases. There is still some acceleration though, so the skydiver's velocity is still getting bigger.

C The velocity, and so the air resistance, increase until the skydiver's weight is balanced by the upward force of air resistance. The resultant force is zero, and the skydiver will not accelerate any more. The velocity of the skydiver stays the same. This steady speed downwards is the **terminal velocity**.

Reducing the terminal velocity

D When the skydiver opens a parachute, the force of air resistance increases. This means that there is a resultant force upwards and an acceleration upwards. The velocity decreases.

As the velocity gets smaller, the air resistance force also decreases. The resultant force upwards is less and the acceleration upwards also decreases.

The resultant upward force and upward acceleration are still there though. The speed still gets less and so the force of air resistance also keeps decreasing.

E The air resistance balances the weight of the skydiver. The resultant force is zero again and the speed downwards now stays the same. This is a new terminal velocity. The new terminal velocity is much lower than the terminal velocity without a parachute.

Shape and terminal velocity

Cars and ships that move horizontally also have a terminal velocity when the driving force from their engine becomes balanced by air resistance.

Falling parachutists want their terminal velocity to be very low. A racing driver wants the terminal velocity of the car to be as high as possible.

The shape of an object affects its terminal velocity. Racing cars are designed to minimise the forces due to air resistance. They have a streamlined shape so that air can flow over them more easily. This means that they can reach a higher terminal velocity than a car that has the same thrust force but meets greater air resistance forces.

◀ This ferry is streamlined

D How does shape affect terminal velocity?

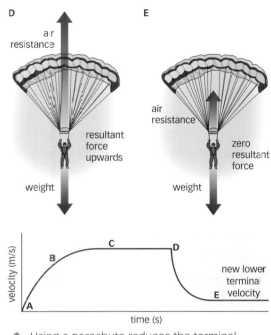
▲ Using a parachute reduces the terminal velocity

Questions

1 A stone is falling through pondwater at a constant velocity. What do you know about the forces acting on the stone? ↓ E

2 Draw a velocity–time graph for a feather that falls off a flying bird.

3 Describe what is happening at each of the labelled points on the velocity–time graphs. Consider the forces on the skydiver at each point. ↓ C

4 How does a parachute reduce a skydiver's terminal velocity?

5 How will the shape of the ferry shown in the photo affect its fuel consumption? ↓ A*

Learning objectives

After studying this topic, you should be able to:

- ✔ describe how forces acting on an object may cause a change in its shape
- ✔ describe the energy changes when a force is used to change the shape of a spring
- ✔ state and use the relationship between the force applied and the extension of the spring

Key words

elastic potential energy, extension, directly proportional, limit of proportionality

A Give an example of an everyday object which stores elastic potential energy.

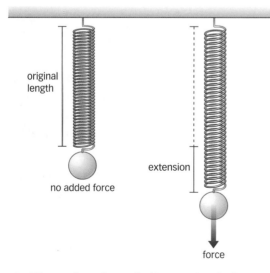

▲ When a force is applied to a spring, it changes shape and extends

Stretching and compressing objects

Forces not only change the way objects are moving, but can also change the shape of objects. They can squash or stretch them.

When a spring is squashed or stretched by a force, energy is transferred into **elastic potential energy** and stored within the spring. When the force is removed, this energy is converted into other forms which return the spring to its original length.

The same thing happens when you pull back an elastic band. This process of storing elastic potential energy is useful for spring toys, wind-up clocks, and car suspension systems.

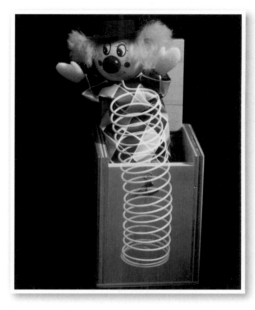

◀ When this toy jumps, it converts elastic potential energy stored in the spring into kinetic energy

Hooke's law

In 1676 the British scientist Robert Hooke investigated the relation between the force applied to an elastic object and the change in its length (called the **extension**).

He found the force applied to an object was **directly proportional** to its extension, up to a certain point. When he doubled the force on a spring he found the extension would also double. If applying 20 N caused a spring to extend 0.05 m, then applying 40 N would cause it to extend 0.10 m.

You may have used a newtonmeter or weighing scales that make use of this relationship. The extension of a calibrated spring is used to determine the weight of an object.

This relationship continues up to the **limit of proportionality**. Above this point, the spring starts to deform – it will no longer return to its original length.

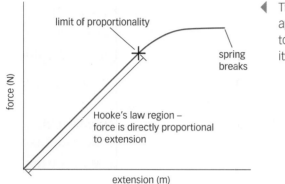

◀ The graph obtained by applying different forces to a spring and measuring its extension

How much a certain spring extends depends on the force applied and the spring constant. This is a measure of the stiffness of the spring, and it is measured in newtons per metre (N/m).

The relationship between force, spring constant and extension is represented in the equation below.

force   =   spring constant   ×   extension
(newtons, N)   (newtons per metre, N/m)   (metres, m)

If the force is called F, the spring constant k, and the extension e, then:

$$F = k \times e$$

Did you know...?

If the force applied to an object is too great, the object might change shape too much, causing it to break. The materials used to build bridges and buildings must be able to withstand the huge forces involved.

Worked example

A spring found in a wind-up toy has a spring constant of 10 N/m. Calculate the force needed to extend the spring by 20 cm.

spring constant = 10 N/m

extension = 20 cm = 0.2 m

$$\text{force} = \frac{\text{spring}}{\text{constant}} \times \text{extension}$$

$$= 10 \text{ N/m} \times 0.2 \text{ m}$$

$$= 2 \text{ N}$$

A much stiffer spring found in a car suspension system has a spring constant of 5000 N/m. Calculate the force needed to extend the spring by 20 cm.

$$\text{force} = \frac{\text{spring}}{\text{constant}} \times \text{extension}$$

$$= 5000 \text{ N/m} \times 0.2 \text{ m}$$

$$= 1000 \text{ N}$$

Questions

1. Describe the energy changes when you compress the spring toy shown in the picture on the left.
   ⬇ E

2. A spring has a spring constant of 2000 N/m and extends 0.5 m when a force is applied. Calculate the value of the force and give the units.
   ⬇ C

3. Describe Hooke's law.

4. A force of 600 N is applied to a spring which extends 20 cm. Find the value of the spring constant.

5. Sketch a graph of force against extension and label the key points. Then draw a second line showing the extension for a spring with a spring constant twice the size of the spring constant of the first spring.
   ⬇ A*

Exam tip

✔ Hooke's law relates the force applied to an object to its extension, not its length. Work out the extension of the object from its original length.

✔ 'Directly proportional' means that if one variable goes up by a factor of x, so does the other variable. If one doubles then the other doubles, or if one goes up by ×5 then the other goes up by ×5.

Learning objectives

After studying this topic, you should be able to:

✔ understand that work done is equal to energy transferred

✔ describe how when a force moves something through a distance, against another force such as gravity or friction, energy is being transferred

✔ use the equation linking the work done, the force and the distance moved in the direction of the force

Key words

work done, joule

▲ Work is done when weights are lifted

A What is the relationship between work done and energy transferred?

B Give three examples of doing work.

Working hard?

In science, the term **work done** (or just 'work') has a very specific meaning. 'Work done' is another way of saying energy has been transferred. Lifting weights, as in the picture on the left, transfers energy to the weights. We can say work has been done on the weights.

Work done   =   energy transferred

When you lift up the weights you might transfer 100 J of energy to them. This means the work done on the weights is 100 J. Work done, just like energy, is measured in **joules**. If no energy is transferred then no work is done.

Other examples of doing work include climbing stairs, pushing a trolley at a supermarket, or pulling a sledge along the ground. Work is always done when a force is used to move something through a distance against an opposing force.

How much work?

The amount of work done on an object depends on the force applied and the distance moved.

$$\text{work done (joules, J)} = \text{force (newtons, N)} \times \text{distance moved in the direction of the force (metres, m)}$$

If W is the work done, F is the force, and d is the distance moved in the direction of the force, then:

$$W = F \times d$$

The distance moved must be in the same direction as the force. On the left of the diagram, the book has been lifted upwards against the gravitational force (weight), which acts downwards. On the right, the book has been lifted sideways and up, but it has been moved the same distance against the gravitational force of weight. The same amount of work has been done against the weight, so the same amount of energy has been transferred to the book.

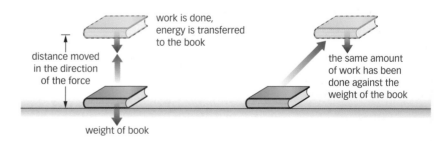

▲ When calculating work done, you use the distance moved in the direction of the force

Worked example

A TV with a weight of 300 N is lifted onto a wall mount 1.2 m from the ground. Calculate the work done.

work done = force × distance moved in the direction of the force,

or $W = F \times d$

force = 300 N

distance moved = 1.2 m

work done = 300 N × 1.2 m

= 360 J

Exam tip

✔ The distance moved must be in the direction of the force. For questions where objects are lifted, this is the vertical distance moved.

✔ Make sure you use the force in the 'work done' equations. You may need to calculate the weight of an object from its mass.

Work done against frictional forces

When objects rub against each other, energy is transferred from kinetic energy into heat. We say that work is being done against friction.

This can be a problem because energy is transferred away from the moving object, but it can be very useful too. Most brakes use this principle to slow vehicles down. Pads or discs rub against moving parts and heat up to very high temperatures. Some kinetic energy is transferred into heat and the vehicle slows down. Next time you stop on your bike, if you put your hand near the brake pad you will be able to feel this heat being transferred to the surroundings.

▲ Friction on a racing car's brakes can make them heat up so much that they glow

▲ The Space Shuttle's kinetic energy is transformed into heat on re-entry

Questions

1 What units are used to measure work?

2 Calculate the work done to pull a sledge 80 m against a frictional force of 6.0 N.

3 Explain why you do more work pulling a sledge uphill than along the flat.

4 A delivery driver lifts 20 boxes, each with a mass of 3.0 kg, into his truck 1.5 m above the ground. Find:

(a) the work done on each box

(b) the energy transferred to each box

(c) the total work done lifting all the boxes into the truck.

Did you know...?

When the Space Shuttle re-enters the Earth's atmosphere it is travelling at a whopping 18 000 mph. At this speed, the air resistance causes the shuttle to heat up to over 1500 °C. Most of the shuttle's kinetic energy is transferred into heat and so it slows down. It lands at just over 200 mph.

Learning objectives

After studying this topic, you should be able to:

✔ state that the unit of power is the watt

✔ describe power as the rate of doing work

✔ use the equation power = work done (or energy transferred) divided by time taken

---

PROline

MOD.:ST44

2450MHz	
230V ∿ 50Hz	MICROWAVE INPUT POWER : 1550 W
	MICROWAVE ENERGY OUTPUT : 950 W

SERIAL NO. 81000138

MADE IN KOREA

CE

WARNING – HIGH VOLTAGE

▲ Different microwaves have different power outputs. This one has a power of 950 W.

Did you know...?

Most humans can produce a power output of just over 100 W, peaking at over 1000 W for very short periods. An average horse can sustain a power output of around 750 W for much longer. Brake horse power (BHP) is often used as a measure of vehicle power output. A family car may have around 100 BHP, or the same power as 100 horses. A high performance car may produce over 500 BHP, that's the same as 500 horses or 375 000 W!

'Watt' is power?

Power, like work, means something different in scientific language from its everyday use. A politician may be a very powerful person in terms of governing a country, but they are likely to be much less powerful physically than a honed athlete.

In the scientific sense of transferring energy, a powerful person or machine can do a lot of work in a short space of time. Power is the work done in a given time (or the rate of doing work).

Boiling water

A more powerful kettle will do more work in a certain time. It will transfer more electrical energy to heat every second. This means the water will boil quicker. A more powerful car transfers more chemical energy (in the fuel) into kinetic energy per second. This means it can accelerate more rapidly to a high speed.

Power is measured in **watts** (W). One watt is one joule of work done (or energy transferred) every second. A 1500 W hairdryer transfers 1500 J of energy every second. A large TV may have a power output of 120 W, an average family car 75 000 W, and an express train a huge 12 000 000 W.

◀ Express trains have a power output that is many times greater than a typical family car

A Name the unit of power.

B How many times more powerful is an express train than a typical family car?

Horses can transfer more energy per second than a human being. They have an average power output of 750 W.

Calculating power

If power is the work done in a given time (rate of doing work), then:

$$\text{power (watts, W)} = \frac{\text{work done (joules, J)}}{\text{time taken (seconds, s)}}$$

'Work done' is just another way of saying 'energy transferred', so this can be written as

$$\text{power (W)} = \frac{\text{energy transferred}}{\text{time taken}} \text{ (J/s)}$$

If power is P, the energy transferred is E, and the time taken for the energy transfer is t, then:

$$P = \frac{E}{t}$$

Worked example 1

A man pushing a wheelbarrow does 400 J of work in 5 s. Calculate the power that he develops.

$$\text{power} = \frac{\text{work done}}{\text{time taken}}$$

work done = 400 J
time taken = 5 s

$$\text{power} = \frac{400\ \text{J}}{5\ \text{s}} = 80\ \text{W}$$

Worked example 2

An electric shower transfers 540 000 J of energy to the water in 1 minute. Calculate its power.

$$\text{power} = \frac{\text{energy transferred}}{\text{time taken}}$$

energy transferred = 540 000 J
time taken = 60 s (as it took 1 minute)

$$\text{power} = \frac{540\,000\ \text{J}}{60\ \text{s}} = 9000\ \text{W or 9 kW}$$

Exam tip

✔ When using the equations for power, you must make sure time is in seconds. If it is in minutes, you will need to convert it to seconds.

Questions

1 What is the equation for power?

2 A cyclist has a power meter on her bike to tell her about her performance. It is showing a steady reading of 200 W. How much work is she doing every second?

3 Calculate the power provided by a small solar panel on a satellite when it transfers 720 J of energy every minute.

4 A boy pulling a sledge does 6000 J of work in 2 minutes. Calculate the power he develops.

5 A 40 W bulb is left on. Calculate the energy transferred in:

(a) 10 seconds

(b) 30 minutes

(c) 24 hours.

14: Gravitational potential energy and kinetic energy

Learning objectives

After studying this topic, you should be able to:

✔ understand what factors affect gravitational potential energy

✔ calculate changes in gravitational potential energy

✔ understand the factors affecting the kinetic energy of an object

✔ use the kinetic energy equation

✔ describe the benefits of regenerative braking

A Apart from gravitational field strength, which two factors affect the GPE of an object?

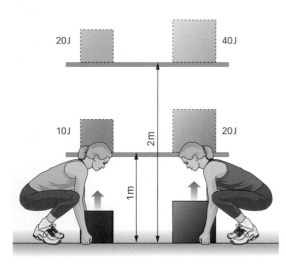

▲ The change in gravitational potential energy (GPE) depends on the mass of the object and the change in height

Gravitational potential energy

When you lift up a book and place it on a shelf, you are doing work on the book. The book gains **gravitational potential energy** (or GPE) as it is lifted away from the ground. The GPE of an object is the energy it has because of its position in a **gravitational field**, like the one around the Earth. This energy depends on

- the mass of the object
- its height above the ground
- the strength of the gravitational field.

The higher you lift an object above the ground, the greater its GPE. To calculate any change in GPE we can use the equation below.

change in GPE (joules, J)	=	mass (kilograms, kg)	×	gravitational field strength (newtons per kilogram, N/kg)	×	change in height (metres, m)

If E_p is the GPE, m the mass, g the gravitational field strength and h the change in height, then:

$$E_p = m \times g \times h$$

On Earth, the gravitational field strength is 10 N/kg.

Worked example 1

Find the change in GPE when a book of mass 1.2 kg is lifted 1.5 m and placed on a shelf.

change in GPE = mass × gravitational field strength × change in height

$$E_p = m \times g \times h$$

mass of book = 1.2 kg, gravitational field strength on Earth = 10 N/kg, and change in height = 1.5 m.

change in GPE = 1.2 kg × 10 N/kg × 1.5 m

= 18 J

B A person of mass 60 kg runs up a flight of stairs 3.0 m high. Calculate their change in GPE.

Kinetic energy

Any moving object has a **kinetic energy** (or KE). The size of the kinetic energy depends on

- the mass of the object
- the speed it is travelling at.

An object with a greater mass will have more kinetic energy. An object that is moving faster will have more kinetic energy. This is the equation that links kinetic energy, mass and speed:

kinetic energy = ½ × mass × speed²
(joules, J)   (kilograms, kg)   (metres per second, m/s)²

If E_k is the kinetic energy, m the mass, and v the speed, then:

$$E_k = ½ \times m \times v^2$$

Key words

gravitational potential energy, kinetic energy, regenerative braking

Worked example 2

Find the kinetic energy of a car of mass 1200 kg travelling at 20 m/s.
kinetic energy = ½ × mass × speed² or $E_k = ½ \times m \times v^2$
mass of the car, m = 1200 kg, and speed of the car, v = 20 m/s
kinetic energy = ½ × 1200 kg × (20 m/s)²
$= ½ \times 1200 \times 400$
= 240 000 J or 240 kJ

Exam tip · AQA

✔ When using the equation for kinetic energy, that is, kinetic energy = ½ × mass × speed², don't forget to square the speed!

Energy transfers

The law of conservation of energy states that energy cannot be created or destroyed. However, it can be transferred. When you drop a ball, its GPE is transferred into kinetic energy as it falls. The same thing happens to bungee jumpers. When they jump, their GPE is converted into kinetic energy as they accelerate towards the ground.

◀ This hybrid car converts kinetic energy into useful electrical energy

Some hybrid cars make use of a clever technology called **regenerative braking**. The brakes in these vehicles are specifically designed to convert some of the kinetic energy back into energy to be stored in the car's batteries. This makes the cars much more fuel-efficient and the brakes wear down less quickly.

Questions

1 State the equation for change in gravitational potential energy and give all the units. ↓ E

2 Calculate the change in gravitational potential energy when a crane lifts a 3000 kg concrete block 40 m into the air. ↓ C

3 Describe an advantage of using regenerative braking.

4 Find the kinetic energy of a football of mass 400 g travelling at 20 m/s

5 Calculate the speed of a horse of mass 600 kg with a kinetic energy of 43 200 J. ↓ A*

Learning objectives

After studying this topic, you should be able to:

- ✔ define and calculate momentum
- ✔ explain and apply the law of conservation of momentum
- ✔ explain the benefit of air bags, crumple zones and other safety devices in cars

▲ The momentum of this runner depends on his velocity and his mass

A A woman with a mass of 60 kg is skydiving and has a velocity of 45 m/s. What is her momentum?

B A football has a mass of 450 g and is moving at 25 m/s. What is its momentum?

Momentum of an object

The **momentum** of a moving object depends on its mass and velocity. You can calculate momentum using the equation:

$$\text{momentum} = \text{mass} \times \text{velocity}$$
$$\text{(kg m/s)} \qquad \text{(kg)} \qquad \text{(m/s)}$$

We can use the symbol p for momentum. Then if mass is m and velocity is v:

$$p = m \times v$$

As with velocity, you have to state the direction of momentum as well as how big it is.

If an object is not moving, its velocity is zero so it does not have any momentum.

Conservation of momentum

The total amount of momentum in a system of objects that interact only with each other always stays the same. For example, if two moving objects collide, the total momentum of the two objects does not change. This is the **law of conservation of momentum**.

Only the objects themselves must be involved. If any external forces act on the objects, such as friction, then momentum is not conserved.

You can use the law of conservation of momentum to solve problems involving collisions and explosions:
total momentum before collision (or explosion) = total momentum after collision (or explosion).

Worked example

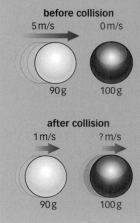

before collision
5 m/s   0 m/s
90 g   100 g

after collision
1 m/s   ? m/s
90 g   100 g

The diagram shows a cue ball of mass 100 g rolling at 5 m/s towards a stationary pool ball. The pool ball has a mass of 160 g. After the collision, the cue ball moves at 1 m/s. What is the velocity of the pool ball?

$$\text{momentum} = \text{mass} \times \text{velocity}$$
$$\frac{\text{total momentum}}{\text{before collision}} = \frac{\text{total momentum}}{\text{after collision}}$$

Before collision:

mass of cue ball = 100 g or 0.1 kg, and its velocity = 5 m/s

mass of pool ball = 160 g or 0.16 kg, and its velocity = 0 m/s (it is not moving)

total momentum = momentum of cue ball + momentum of pool ball

= (0.1 kg × 5 m/s)   + (0.16 kg × 0 m/s)

= 0.5 kg m/s

After collision:

velocity of cue ball = 1 m/s (It is moving in the same direction as it was originally, so the sign is +. If it was moving in the opposite direction, it would be −1 m/s.)

velocity of pool ball = V m/s

total momentum = momentum of cue ball + momentum of pool ball

= (0.1 kg × 1 m/s)   + (0.16 kg × V m/s)

= 0.1 kg m/s   + (0.16 × V) kg m/s

Total momentum before collision = total momentum after collision

0.5 kg m/s   = 0.1 kg m/s + (0.16 × V) kg m/s

Rearranging, (0.16 × V) kg m/s = (0.5 − 0.1) kg m/s = 0.4 kg m/s

$$V = \frac{0.4}{0.16} \text{ m/s}$$

After the collision, the velocity V of the pool ball is 2.5 m/s.

Car safety

In a crash, a car stops very quickly. Both the car and the occupants have kinetic energy and momentum. You will keep moving and will hit parts of the car such as the windscreen unless you are wearing a seatbelt.

The sudden stop means that a very large change in momentum happens very quickly. The longer the time that this takes to happen, the smaller the forces involved. Air bags, and stretching seat belts increase the time taken for you to stop and crumple zones mean the car takes longer to stop, significantly reducing the forces acting.

These features and others such as side impact bars also absorb some of the huge amounts of kinetic energy involved when they deform.

Key words

momentum, law of conservation of momentum

Exam tip

✔ Remember that mass must be in kilograms not grams.

Questions

1 What is the momentum of an object?

2 Look at the two cars in the diagram.
   (a) Calculate the momentum of each car.
   (b) Explain why they are different.

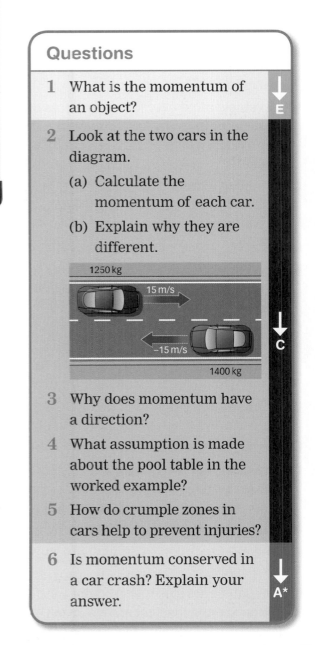

3 Why does momentum have a direction?

4 What assumption is made about the pool table in the worked example?

5 How do crumple zones in cars help to prevent injuries?

6 Is momentum conserved in a car crash? Explain your answer.

Course catch-up

Revision checklist

- Forces are pushes and pulls. Objects exert equal and opposite forces on one another. Resultant force is the sum of all the forces acting on an object.
- Resultant forces change the motion of objects, causing them to accelerate. Resultant force = mass × acceleration.
- Speed is a measure of how fast something is moving. Speed = distance/time. This principle is used in speed cameras.
- Velocity describes direction as well as speed.
- The gradient of a distance–time graph represents speed. Acceleration is a change in speed, and can be positive (speeding up) or negative (slowing down).
- The gradient of a velocity–time graph represents acceleration. The area under a velocity–time graph represents distance travelled.
- A vehicle must overcome the forces of friction and air resistance in order to start moving. When these are balanced by the vehicle's driving force, the vehicle is travelling at a steady speed.
- Stopping distance (thinking distance plus braking distance) is affected by speed, road conditions, vehicle condition, and driver reaction time.
- Weight = mass × gravitational field strength.
- As an object falls under gravity, its velocity increases until its weight is balanced by the force of air resistance. The object then reaches a terminal velocity (steady speed).
- Forces can change the shape of objects. Squashing or stretching a spring gives it elastic potential energy. Hooke's law dictates that the force applied to an object is directly proportional to its extension, up to a certain point.
- Work done is calculated using the equation, work done = force × distance moved in the direction of the force.
- Power = work done/time taken.
- Gravitational potential energy (GPE) is defined by mass, height above the ground, and the Earth's gravitational field. GPE is transferred into kinetic energy (KE) when an object moves.
- A moving object's KE is defined by its mass and its speed.
- The momentum of a moving object is defined by its mass and its velocity.
- Car safety devices increase the time taken for the change in momentum experienced in a crash.

regenerative braking uses kinetic energy to recharge battery

brakes convert kinetic energy to heat

stopping distance = thinking distance + braking distance

poor road and vehicle conditions increase braking distance

speed (metres per second, m/s)

cameras

safety features reduce momentum gradually

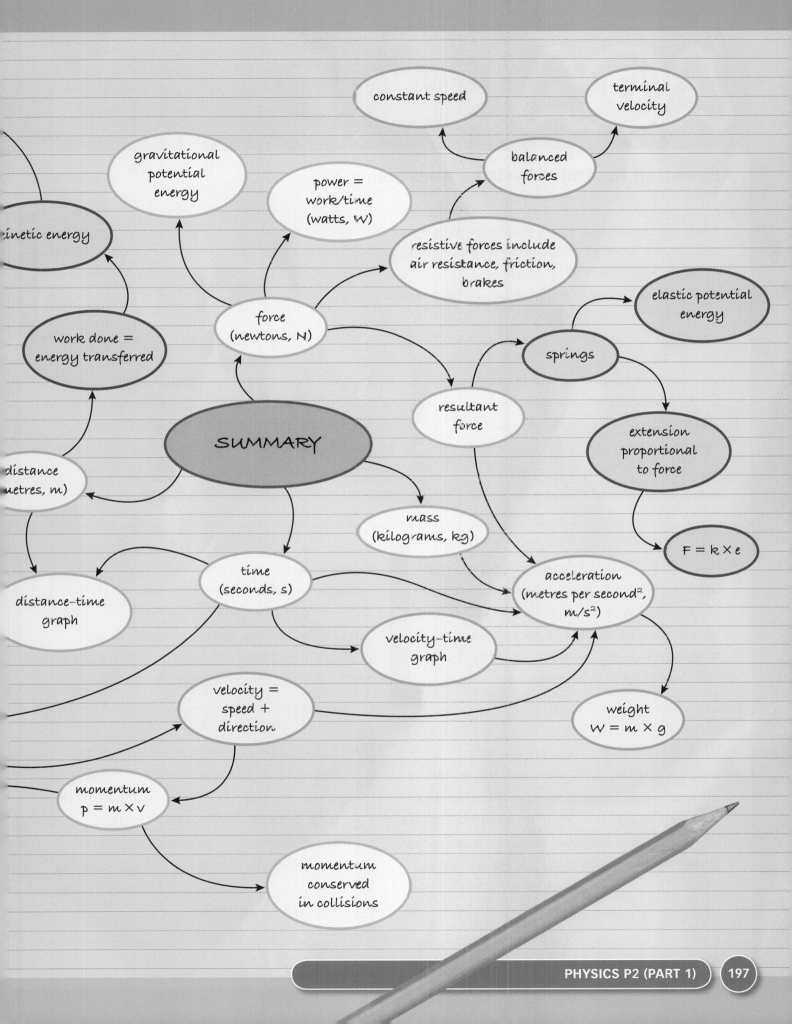

constant speed

terminal velocity

gravitational potential energy

power = work/time (watts, W)

balanced forces

...netic energy

resistive forces include air resistance, friction, brakes

work done = energy transferred

force (newtons, N)

elastic potential energy

springs

resultant force

SUMMARY

extension proportional to force

...distance (metres, m)

mass (kilograms, kg)

$F = k \times e$

distance-time graph

time (seconds, s)

acceleration (metres per second^2, m/s^2)

velocity-time graph

velocity = speed + direction

weight $W = m \times g$

momentum $p = m \times v$

momentum conserved in collisions

Answering Extended Writing questions

The graph shows the speed of fall of a skydiver at various stages of a descent. Explain the motion of the parachutist at each of the stages A–E on the graph.

The quality of written communication will be assessed in your answer to this question.

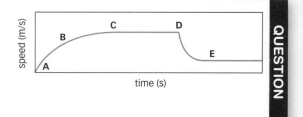

G–E

At A she has jumpd out and is speeding up At B are resistance is slowwing her down, C is terminal spede. At D she opens the parachute, goes up for a bit till her chute is fuly open and at E she comes down a good desent and lands.

Examiner: This answer shows only vague understanding of the physics, though some words are used correctly. There is again almost no mention of forces, and the actual motion at D is misunderstood. There is some irrelevant information. Spelling, punctuation, and grammar are erratic.

D–C

At A she has just left the aeroplane, speeding up. At B there is air resistence, so she slows down a bit. At C she has reached terminal speed. She opens her parachute at D, slows down. By E she has reached terminal speed with her parachute open.

Examiner: Most but not all of the physics is correct – at B she is not slowing down (though this is a common mistake). Answer includes little about forces acting, which is crucial to the explanations. There are occasional errors in spelling, punctuation, and grammar.

B–A*

At A she is accelerating quickly, as air resistance is low. At B she is moving faster, so air resistance is higher, so resultant force and acceleration are lower. At C air resistance equals weight, she is at terminal speed. At D she opens parachute – area and so air resistance increase greatly, so she decelerates. At E she has reached new lower terminal speed.

Examiner: This answer refers to both forces and consequent acceleration at each stage – sometimes by implication. No link is made between acceleration and the gradient of the graph, though this is implied. The physics explanations and the use of words are all correct. Spelling, punctuation, and grammar are all good.

Exam-style questions

1 Match these quantities with their units.

A01

force	m/s
mass	m
acceleration	J
velocity	m/s²
kinetic energy	W
distance	kg
power	N

2 This is a velocity–time graph for a tube train moving between stations.

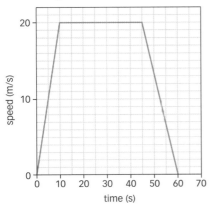

a How long did the journey take altogether?

b Calculate the acceleration in the time 0–10 s.

c Calculate the acceleration in the time 45 s–60 s.

d Describe what is happening in the time 45 s–60 s.

e Calculate the distance the train moves while it is travelling at a steady speed.

3 A weightlifter heaves a load of 30 kg from the floor to a height of 2 m. He then drops the bar, which hits the ground.

a Calculate the weight of the load.

b How much gravitational potential energy does the load have at the top of the lift?

c Calculate the speed at which the load hits the floor.

4 A model rocket has mass 5 kg. When it is fired vertically upwards, the initial thrust from the exhaust gas is 80 N.

a What is the resultant upward force accelerating the rocket?

b What is the initial acceleration of the rocket?

c Assuming the thrust remains 80 N, explain why the acceleration of the rocket would increase.

Extended Writing

5 What is meant by the terms stopping distance, thinking distance, and braking distance?
Explain why a car's braking distance is greater if the road is icy or wet.

6 An archer fits an arrow to his bow, draws the bow back and then fires the arrow, which sticks into a target. Describe the energy changes that occur.

7 Explain how air bags in a car help to protect you if a collision occurs.

P2 Part 2

Electricity and radiation

Why study this unit?

What would the world be like without electricity? A flow of tiny, negatively charged, sub-atomic particles is vital to the operation of every electrical appliance, from small touch-screen mobile phones to large 3D TVs. An understanding of electric current is essential to all scientists, engineers, and anyone interested in how things work. In this unit you will learn about electric circuits, the differences between static, current, and mains electricity, how to control an electric current, and some of the dangers of all forms of electricity.

You will also learn more about the atom, and how radiation from unstable nuclei breaking down surrounds us all of the time, continuously bombarding the cells in our bodies. Finally, you will learn how scientists have been able to split the atom to devastating effect, while they have yet to master fusing it back together. Looking up at the stars may yet offer solutions to this challenge, and provide a clean, cheap energy source for the future.

You should remember

1 Electricity can produce a variety of different effects, including heating.

2 There are two different types of electric circuit: series and parallel.

3 Electric circuits can be used to control an electric current.

4 Some materials are charged. Opposite charges attract, and like charges repel.

5 All materials are made up of atoms.

Lightning is one of nature's most impressive and terrifying uses of electricity. Each strike only lasts around 30 millionths of a second, but in that time it transfers a massive 5 billion joules of energy to its surroundings. The voltage in a strike can be as high as 100 million volts - that's the same as 66 million AA batteries.

Even more frightening is what happens to the air around this giant electric spark. The air is heated to almost 30 000 °C, which is five times hotter than the surface of the Sun. The heating takes place so rapidly that the air expands in a supersonic shock wave. We hear this explosion as thunder.

Learning objectives

After studying this topic, you should be able to:

- ✔ describe how some materials can become charged by rubbing them
- ✔ describe static electricity effects in terms of the transfer of electrons
- ✔ explain that like charges repel and opposite charges attract

A Why might rubbing an insulator cause it to become charged?

B Has a positively charged object gained or lost electrons?

Exam tip AQA

- ✔ Remember that you can only charge something with static electricity by rubbing two insulators together.

Key words

static electricity, insulating material, electrostatic charge, repel, attract

Electrostatic charge

You have experienced an effect of **static electricity** if you have ever had a shock when touching a metal door knob, or getting out of a car.

When you rub certain types of **insulating materials** together, they can become charged. This is sometimes known as an **electrostatic charge**.

Insulating materials can be charged with positive charge or negative charge. The charge is caused by electrons, which have a negative charge, being transferred from one insulating material to another when they are rubbed together.

Positives and negatives

When an acetate rod is rubbed with a woollen cloth, electrons move from the acetate rod to the cloth. The acetate rod becomes positively charged.

When a polythene rod is rubbed with a woollen cloth, electrons move from the cloth to the polythene rod. The polythene rod becomes negatively charged.

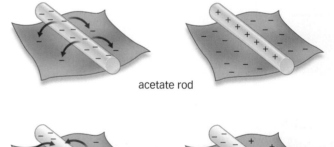

acetate rod

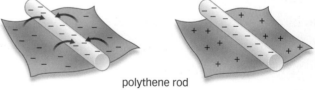

polythene rod

▲ The rods become positively or negatively charged when electrons are transferred

These effects are only seen with insulating materials, where the transferred charge will stay in a particular area. In materials such as metals, electrons can move freely through the material and can travel to earth, charge does not accumulate in one place.

Opposite charges attract, like charges repel

If two bodies are both positively charged, they will **repel** each other, or push each other away. If both bodies are negatively charged, they will also repel each other. But if the two bodies have opposite charges, they will **attract** each other.

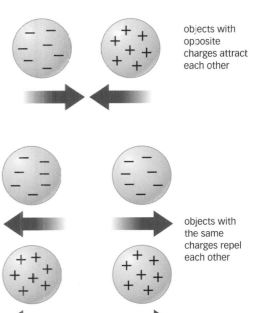

objects with opposite charges attract each other

objects with the same charges repel each other

▲ Attraction and repulsion of charged particles

▲ The uncharged stream of water is attracted to the charged balloon

Questions

1 Name the two types of charge.

2 What kind of material can become charged when you rub it?

3 Do like charges attract or repel each other?

4 Explain how a plastic comb could be used to pick up small pieces of paper.

5 (a) Explain how an acetate rod becomes charged when you rub it with a woollen cloth.

   (b) A polythene rod is rubbed with metal foil. Will it become charged? Explain your answer.

Did you know...?

Charged objects can attract other uncharged objects. The balloon in the photo on the left has been negatively charged by rubbing. The negative charge on the balloon is repelling away electrons in the stream of water. So the surface of the water nearest the balloon is now positively charged. The stream of water is attracted towards the balloon.

▲ The boy has gained an electrostatic charge while sliding down. His hair is sticking out because the charges on the hairs are all the same type, so they are repelling one another.

Learning objectives

After studying this topic, you should be able to:

✔ understand that electric current is the rate of flow of charge

✔ calculate the size of an electric current

✔ explain what potential difference is

✔ explain how to connect cells in series to increase the potential difference

---

A What type of charge do electrons carry?

B A charge of 1950 C flows past a point in a kettle element in 150 seconds. What is the size of the current flowing through the kettle?

Current and charge

An electric **current** is a flow of **charge** around a circuit. Negatively charged electrons flow from the negative terminal of a **cell** round the circuit to the positive terminal.

The charge carried by each electron is very small, so electric charge is measured in a much larger unit called the **coulomb**, with the symbol C.

The size of the current is given by the amount of charge passing a point in a circuit each second.

You can calculate the size of the current flowing in a circuit by using the equation:

$$\frac{\text{current}}{\text{(amperes, A)}} = \frac{\text{charge flowing past a point (coulombs, C)}}{\text{time (seconds, s)}}$$

If the current is called I, the charge Q, and the time t, then:

$$I = \frac{Q}{t}$$

Worked example 1

In 60 seconds a charge of 300 C flows past a point in a circuit. What is the current in the circuit?

$$\text{current} = \frac{\text{charge flowing past a point}}{\text{time}} \text{ or } I = \frac{Q}{t}$$

charge = 300 C and time = 60 s

$$\text{current} = \frac{300 \text{ C}}{60 \text{ s}}$$

$$= 5 \text{ A}$$

Nowadays, we know that the negatively charged electrons move from the negative terminal of a cell through the circuit to the positive terminal. However, earlier scientists thought that charge moved from the positive terminal to the negative; they showed this **conventional current** direction on their circuit diagrams. We still use this notation today – circuit diagrams show current flowing from positive to negative.

Potential difference

As electrons pass through the cell in a circuit, they gain energy. They lose energy as they pass through the components of the circuit, when their electrical energy is transferred into other types of energy.

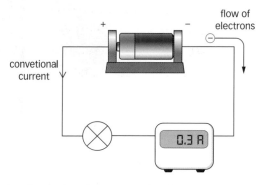

▲ Circuit showing conventional current and electron flow

The difference between the energy carried by the current going into a component and the energy carried by the current leaving the component is measured using the **potential difference**.

The potential difference is the amount of energy transferred (or work done) for each coulomb of charge as it passes through the component. It is measured in volts.

$$\text{potential difference (volts, V)} = \frac{\text{energy transferred (or work done) (joules, J)}}{\text{charge (coulombs, C)}}$$

If the potential difference is called V, the work done (or energy transferred) W and the time t, then:

$$V = \frac{W}{Q}$$

Worked example 2

When a charge of 30 C flows through a DVD player, 6900 J of electrical energy are transferred. What is the potential difference across the DVD player?

$$\text{potential difference} = \frac{\text{energy transferred}}{\text{charge}}$$

energy transferred = 6900 J and charge = 30 C

$$\text{potential difference} = \frac{6900 \text{ J}}{30 \text{ C}}$$

$$= 230 \text{ V}$$

Potential difference is also known as **voltage**. It can be measured using a voltmeter that is connected across a component.

Increasing potential difference

A cell usually supplies a potential difference or voltage of about 1.5 V. Many appliances need a higher potential difference than this. You can increase the potential difference supplied by connecting cells end to end in **series** to form a **battery**. We can roughly think of each cell providing a 'push' to the electrons.

You can find the total potential difference by adding up the potential differences of all the connected cells. But the cells must all be 'pushing' the same way – the positive terminal of one cell must be connected to the negative terminal of the next to increase the potential difference.

Key words

current, charge, cell, coulomb, conventional current, potential difference, voltage, series, battery

Did you know...?

It takes the charge on 6 241 509 750 000 000 000 electrons to provide one coulomb of charge!

C When a car engine is started, 600 J is transferred for a charge of 50 C. What is the potential difference across the engine?

Questions

1 What is an electric current?

2 A torch needs 4.5 V. How many 1.5 V cells do you need to use?

3 A charge of 1000 C flows past a point in a hairdryer circuit in 250 seconds. What is the current through the hairdryer?

4 A lamp transfers 12 000 J of energy as a charge of 500 C passes through it. What is the potential difference across the lamp?

5 A current of 0.25 A flows through a computer for 30 minutes. How much charge passes through the computer?

Learning objectives

After studying this topic, you should be able to:

✔ recognise standard circuit symbols
✔ draw and interpret circuit diagrams

Did you know...?

Circuit symbols are now an almost universal language. It means that when you draw a circuit diagram, most electricians around the world can understand it.

Circuit symbols

These are some circuit components and their symbols.

Symbol	Component	Image
	open **switch** / closed switch	
	cell	
	battery	
	diode	
	resistor	
	variable resistor	
	lamp	
	fuse	
V A	voltmeter / ammeter	
	thermistor	
	light-dependent resistor (LDR)	
	light-emitting diode (LED)	

Exam tip   **AQA**

✔ You need to know all of the standard symbols shown in the table.

A Draw the symbols for a variable resistor, a light-emitting diode (LED), an ammeter, and a voltmeter.

Circuit diagrams

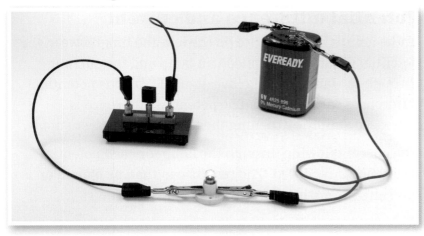

▲ A simple circuit with a battery, lamp, and switch

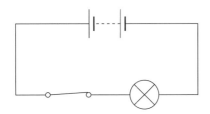

▲ The circuit diagram for the circuit shown in the photo

Look at the circuit shown in the photo. You can draw a circuit diagram using the standard symbols.

You can use circuit diagrams to set up or describe a circuit, or to interpret how a particular circuit might behave.

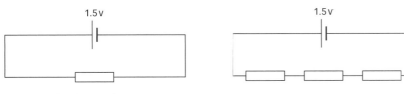

▲ Two simple circuits

All the resistors in the two circuit diagrams above are identical. So it will take charged electrons longer to move around the circuit with three resistors. The flow of charge is slower. In other words, the current is smaller.

> B What components are shown in these two circuit diagrams?
>
> C If all four resistors are identical, which circuit has the highest current?

Key words

switch, diode, resistor, variable resistor, fuse, thermistor, light-dependent resistor, light-emitting diode

Questions

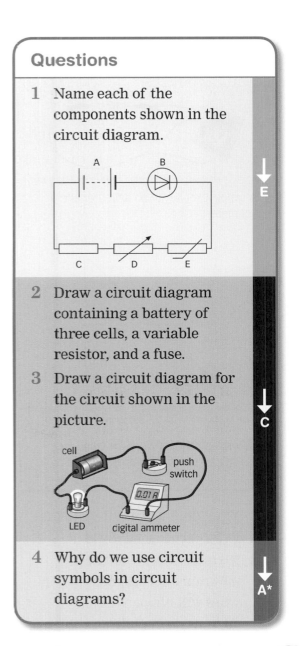

1 Name each of the components shown in the circuit diagram.

2 Draw a circuit diagram containing a battery of three cells, a variable resistor, and a fuse.

3 Draw a circuit diagram for the circuit shown in the picture.

cell
push switch
0.01 A
LED   digital ammeter

4 Why do we use circuit symbols in circuit diagrams?

E

C

A*

Learning objectives

After studying this topic, you should be able to:

✔ explain what current–potential difference graphs show

✔ describe how to find the resistance of a component

✔ calculate current, potential difference, and resistance

▲ A simple circuit with an ammeter and a voltmeter

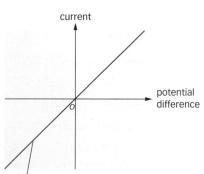

This part of the graph shows that if you reverse the potential difference, you also reverse the direction of the current

▲ Current–potential difference graph for a resistor (at constant temperature)

A What does a current–potential difference graph show?

B If you halve the potential difference across the resistor, what will happen to the current through it?

Potential difference and current

In the circuit in the picture on the left, the link between the potential difference supplied to a lamp and the current through the lamp is being investigated. Various potential differences can be set up. The potential difference is increased simply by turning the knob on the power supply.

The current flowing through the lamp for each potential difference is recorded. The results can then be plotted on a **current–potential difference graph**.

You can replace the lamp in this circuit with different components, and plot the graph of current versus potential difference for each one.

Current–potential difference graph for a resistor

In the graph on the left the line passes through zero and has a straight upward slope (or constant positive gradient). This shows that:

- The current through the resistor is directly proportional to the potential difference across the resistor.
- If you double the potential difference, the current will also double.
- Any change in the potential difference changes the current in the same proportion.

The graph for a resistor will only be a straight line if the temperature of the resistor stays the same.

Calculating resistance, current, and potential difference

If it is more difficult for electrons to pass through a component, we say that the component has a higher **resistance**. This value is shown by the symbol R, and the units are 'ohms' (symbol Ω).

Resistance is linked to current and potential difference by the equation:

$$\underset{\text{(volts, V)}}{\text{potential difference}} = \underset{\text{(amperes, A)}}{\text{current}} \times \underset{\text{(ohms, }\Omega\text{)}}{\text{resistance}}$$

If potential difference is V, current is I and resistance is R, then:

$$V = I \times R$$

Worked example

The potential difference across a lamp is 12 V, and the current flowing through it is 6 A. What is the resistance of the lamp?

potential difference = current × resistance or $V = IR$

$$\text{resistance } (\Omega) = \frac{\text{potential difference (V)}}{\text{current (A)}} = \frac{12\,\text{V}}{6\,\text{A}} = 2\,\Omega$$

Key words

current–potential difference graph, resistance

Resistors often have resistance values that are thousands of ohms. So resistance is often given in kilo-ohms (kΩ), with 1 kΩ = 1000 Ω. When the resistance is large, the current will be small – much less than 1 A. When this happens, currents are given in milliamps (mA), with 1 A = 1000 mA.

Exam tip

✔ Take care with the units of current and resistance when using them in calculations. Always make sure you convert kilo-ohms to ohms and milliamps to amps when needed.

✔ Remember that when resistance goes up, current comes down.

Resistance and current

You can see from the current–potential difference graphs below that for a potential difference of 10 V, the current through resistor A is 0.3 A, but the current through resistor C is only 0.1 A. It is more difficult for charge to pass through resistor C: it has the highest resistance of the three resistors.

At a particular value of potential difference, the current through any component depends on the resistance of the component. If the resistance increases, the current through the component decreases.

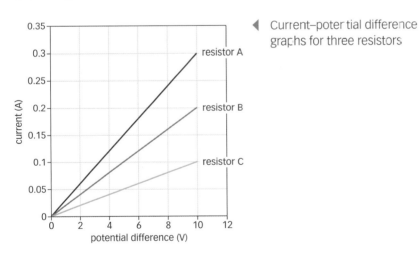

◀ Current–potential difference graphs for three resistors

Questions

1 A current–potential difference graph is a straight line. What does this tell you?

2 The current through a 8 Ω resistor is 1.5 A. What is the potential difference across the resistor?

3 Use points from the graph to calculate the resistances of resistors A, B and C.

4 The potential difference across a variable resistor is 12 V. Calculate the current through the resistor for the following resistances:

(a) 6 Ω

(b) 24 Ω

(c) 1.2 kΩ.

5 What does the gradient of a current–potential difference graph at a particular point represent?

C How do you know that resistor C has the highest resistance?

Learning objectives

After studying this topic, you should be able to:

- ✔ calculate the resistance of components connected in series
- ✔ explain that the size of the current is the same throughout a series circuit
- ✔ describe how the potential difference provided by the supply is shared between the components

Did you know...?

Some Christmas tree lights are connected in series – if one of the bulbs breaks, none of the lights will work.

Series connection

In the circuit shown in the picture, all the electrons moving round must pass through first one lamp then the other. There is only one possible path that the current can take. The lamps are connected in series.

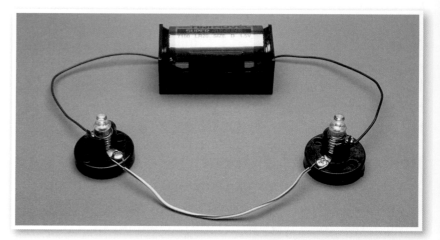

▲ These lamps are connected in series

Resistance in a series circuit

Often, several components can be connected in series in a circuit. It is harder for the current to flow through these circuits because electrons have to flow through each component in turn.

If you add more components, the resistance of the circuit will increase. All components have a resistance, including lamps, LDRs, and thermistors.

You can find the total resistance of the circuit by adding together the resistances of all the components.

A Three resistors with resistances 5 Ω, 10 Ω, and 20 Ω are connected in series. What is their total resistance?

Worked example

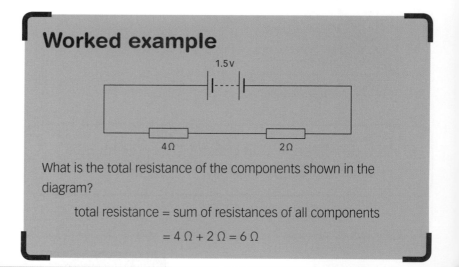

What is the total resistance of the components shown in the diagram?

total resistance = sum of resistances of all components

$$= 4\,\Omega + 2\,\Omega = 6\,\Omega$$

Current in a series circuit

The current in a series circuit is the same everywhere in the circuit. In the circuit shown here, all three ammeters show that the current flowing in the circuit is 0.3 A.

The current is the same throughout the circuit because the electrons have to flow at the same rate through all the components in the circuit.

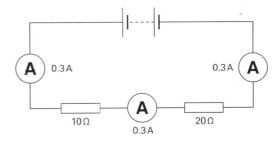

▲ The same current flows through every component in a series circuit

Potential difference in a series circuit

The potential difference of the supply is shared between the resistors in the circuit. The sum of the potential differences across each component will equal the potential difference of the supply.

The total potential difference across the two resistors is 9 V. The potential difference across the 10 Ω resistor is 3 V and the potential difference across the 20 Ω resistor is 6 V. The potential difference has been shared between the components in proportion to the size of the resistance.

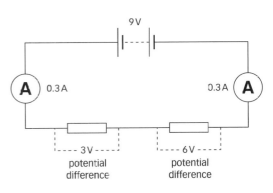
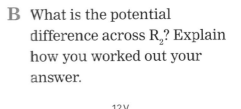

▲ The potential difference of the supply is shared between the resistors

Questions

1 Why is the current the same everywhere in a series circuit?

2 What can we say about the potential difference across components in a series circuit?

3 In the worked example circuit, what is the current if the potential difference of the supply is 3 V?

4 Look at the two circuit diagrams.

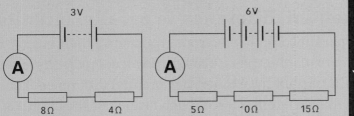

(a) What is the total resistance of each circuit?

(b) What is the current flowing in each circuit?

(c) Calculate the potential difference across each resistor.

5 Should an ammeter have a high or low resistance? Explain your answer.

B What is the potential difference across R₂? Explain how you worked out your answer.

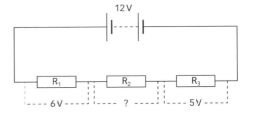

Exam tip AQA

✓ Remember that an ammeter is always connected in series.

Learning objectives

After studying this topic, you should be able to:

✔ explain that the potential difference across components connected in parallel is the same

✔ describe how the current in a parallel circuit splits between the branches

Key words

parallel circuit

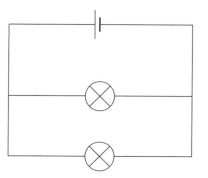

▲ Circuit diagram for the lamps connected in parallel

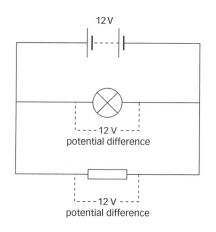

12 V

- - - - 12 V - - - -
potential difference

- - - - 12 V - - - -
potential difference

▲ The potential difference is the same across parallel branches in a circuit

Parallel connections

In the circuit shown below, there is more than one route that electrons can take when moving round the circuit. They can move through one lamp or the other. There are two branches that the current can flow through. The lamps are connected in parallel. A **parallel circuit** has two or more branches.

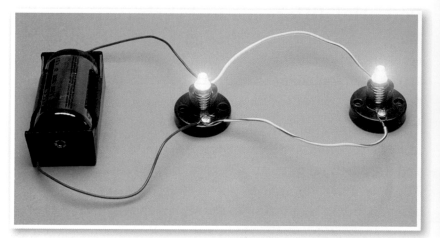

▲ These lamps are connected in parallel

Resistance in parallel circuits

The resistance in a parallel circuit is lower, because the electrons have alternative ways to flow round the circuit. All the charge does not have to flow through all the components.

A Why is resistance lower in parallel circuits?

Potential difference in parallel circuits

In a parallel circuit, electrons transfer all their energy to the components they pass through. So all components in a parallel circuit have the same potential difference. The potential difference across each branch of the circuit is the same because it does not matter which branch the electrons travel through. Each branch is exposed to the full push of the power supply.

B Why is the potential difference the same across the parallel branches in a circuit?

Current in parallel circuits

As there is more than one route for the current to take around a parallel circuit, the current splits between the different branches.

The sum of the currents flowing through all the branches is equal to the total current leaving the battery.

In the circuit shown in the diagram on the right, the current splits into two at **X**. You can see that 0.1 A goes through the lamp, and 0.3 A goes through the resistor. The total current is 0.3 A + 0.1 A = 0.4 A.

At **Y**, the current flowing from the lamp (0.1 A) joins with the current from the resistor (0.3 A) to form a current of 0.4 A again.

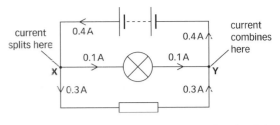

▲ The current from the main part of the circuit is shared between the parallel branches

C What is the sum of the current flowing through the two branches?

Questions

1 What is a parallel circuit?

2 Look at the circuit diagram.

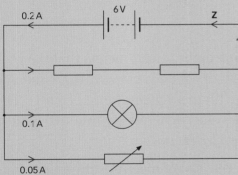

(a) What current is flowing through the branch containing the two resistors? Explain how you worked out your answer.

(b) What is the size of the current at **Z**?

(c) The variable resistance is increased. How will this affect the current flowing through it?

(d) How will it affect the current flowing in the other branches?

(e) What effect would reducing the value of the variable resistance to zero have?

3 Should a voltmeter have a high or a low resistance? Explain your answer.

Did you know...?

When you measure the potential difference across a component, you always connect the voltmeter in parallel.

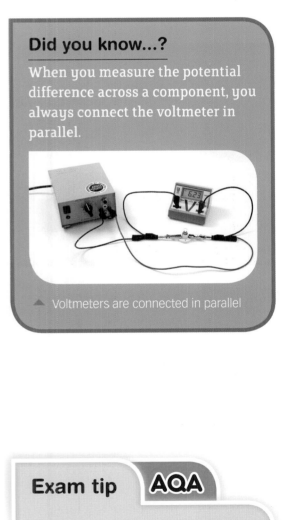

▲ Voltmeters are connected in parallel

Exam tip   AQA

✔ Remember that there is more than one path for the current to take in a parallel circuit.

Learning objectives

After studying this topic, you should be able to:

✔ describe how the resistance of a filament lamp varies

✔ describe what a diode does

✔ understand that a light-emitting diode (LED) emits light when a current flows through it in one direction

✔ compare the uses of different forms of lighting

▲ A filament lamp

Why does resistance increase with temperature in a filament lamp?

The atoms in a conductor are positively charged ions with electrons that are free to move through the metal. As the temperature increases, the ions vibrate more. This makes it more difficult for the electrons to move through the metal.

Resistance of a lamp

A **filament lamp** consists of a thin coil of wire which is usually made of a metal with a high melting point such as tungsten. When a current flows through the wire, it glows brightly and becomes hot.

When you plot a graph of current against potential difference for a filament lamp, you find that it is not a straight line. It is a curve.

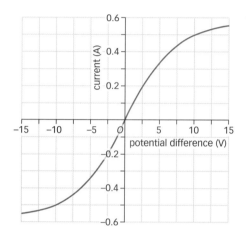

◀ Current–potential difference graph for a filament lamp

As the graph is not a straight line, it means that the resistance is not constant – it does not stay the same. As the current increases, the filament glows more brightly. The filament also transfers electrical energy into thermal energy, so its temperature increases.

An electric current is the flow of electrons through a material. As the temperature of the coil in the filament lamp increases, it is more difficult for the electrons to flow through the metal. The resistance increases as the temperature increases.

Diodes

A diode is a device which lets the current flow through it in one direction only, the 'forward' direction.

At low potential differences, only a very small current flows – typically about 0.001 mA. When the potential difference increases to about 1 V, the resistance of the diode decreases and a current starts to flow through it.

If a diode is connected the other way round, it will allow only a very small current to flow, which is almost zero. This means that the diode has a very high resistance when it is connected in the reverse direction.

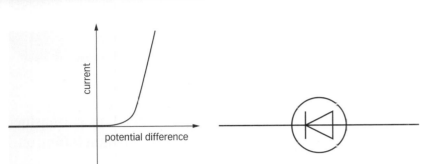

▲ Current–potential difference graph for a diode

▲ Current flows through a diode in the direction of the arrow in the symbol, from positive to negative

LEDs and other types of lighting

An LED, or light-emitting diode, produces light when a current flows through it.

LEDs have a high efficiency and also last a long time. They use a much smaller current than other forms of lighting. They are beginning to be used as replacements for conventional filament light bulbs.

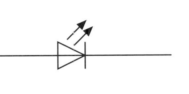

▲ An LED and its circuit symbol

	Type of bulb			
	Filament	Compact fluorescent	Halogen	LED
Power rating	105 W	12 W	20 W	4 W
Efficiency	7.5%	45%	15%	75%
Lifetime	2000 hours	8000 hours	1000 hours	20 000 hours
Cost	£2	£2	£2	£6

A What is an LED?

B Which type of light bulb is the most efficient?

Key words

filament lamp

Exam tip

✓ Remember that when a graph of current against potential difference is a straight line, resistance is constant.

Questions

1 What happens to the resistance of a filament lamp as the temperature increases?

2 From the graph on the previous page, calculate the resistance of the filament lamp at:

(a) 5 V   (b) 10 V.

What do you notice about the resistance?

3 (a) How much useful light power is produced by each type of bulb?

(b) Which type of light bulb has the lowest cost for 20 000 hours of use?

(c) Which type of light bulb would you recommend using? Explain why.

4 When a diode is connected in reverse, a very small current of about 0.001 mA flows. Calculate the resistance of the diode if the potential difference is 2 V.

E

C

A*

23: Light-dependent resistors (LDRs) and thermistors

Learning objectives

After studying this topic, you should be able to:

- ✔ understand how a light-dependent resistor (LDR) works
- ✔ understand how a thermistor works
- ✔ describe applications of LDRs and thermistors

Light-dependent resistors

A light-dependent resistor (LDR) is a special type of resistor. Its resistance changes as the intensity of the light falling on it changes.

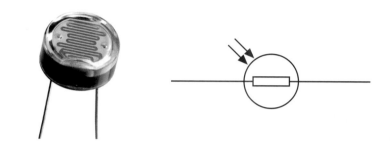

▲ An LDR and its circuit symbol

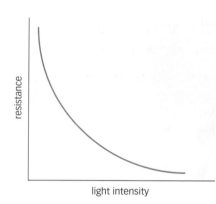

▲ How the resistance of an LDR (light-dependent resistor) varies with intensity of light

When the light levels are low, the resistance of an LDR is high. This is because an LDR is made of a semiconductor material where the outer electrons are bound weakly to the atoms.

When bright light shines on an LDR, the resistance is much lower. The light energy is transferred to the outer electrons which can then break free from the atoms. They are then free to flow through the LDR.

This change in resistance according to the intensity of light means that LDRs can be used as switches. For example, they can be used to switch on security lights when it gets dark.

A What happens to the resistance of an LDR as the intensity of light increases?

Did you know...?

Many street lamps are controlled by light-dependent resistors. If dark clouds reduce the light levels enough, the resistances of the LDRs increase and the street lamps switch on.

Thermistors

A thermistor is another special type of resistor. Its resistance changes as its temperature changes.

When the temperature of the thermistor is low, its resistance is high. This is because a thermistor is made of a material which does not conduct electricity well at low temperatures. The outer electrons are loosely bound to the atoms, and are not free to flow through the thermistor.

As the temperature increases, more outer electrons gain enough energy to break free from atoms. The electrons are then free to flow through the thermistor. So as the temperature increases, the resistance of a thermistor decreases.

Thermistors can be used as temperature sensors. For example, thermistors are often used in car engines to monitor the temperature of the cooling system. They can be used to switch on a fan if the cooling system goes above a certain temperature. They can also be used to warn the driver if the cooling system is about to overheat.

Thermistors are typically used to measure temperatures in the range –90 °C to 200 °C.

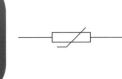

▲ A thermistor and its circuit symbol

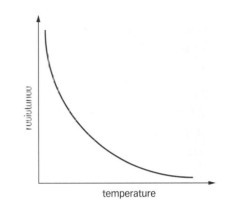

▲ How the resistance of a thermistor varies with temperature

B What happens to the resistance of a thermistor as the temperature increases?

▲ Many digital thermometers use a thermistor for measuring the temperature

Questions

1 (a) What is a thermistor?

   (b) What is an LDR?

2 Describe what will happen to the current in the circuit shown in the diagram as the intensity of light increases.

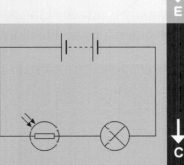

   A circuit with an LDR in it ▶

3 Describe how a thermistor could be used to control the temperature in a personal computer.

4 What other applications might LDRs and thermistors be used for?

5 Draw a circuit diagram for a warning system to alert a driver that a car engine is getting too hot.

↓ E

↓ C

↓ A*

Exam tip   AQA

✔ With a light-dependent resistor, remember that when the light is brighter, the resistance is lower. Similarly, with a thermistor, remember that when the temperature is higher, the resistance is lower.

Learning objectives

After studying this topic, you should be able to:

- ✔ describe electric current as either direct current or alternating current
- ✔ know that mains electricity in the UK is an a.c. supply
- ✔ state that mains electricity in the UK is 230 V a.c. and 50 Hz
- ✔ determine the frequency of an a.c. supply from an oscilloscope

Direct vs alternating current

Think of how many electrical appliances you use. They all require an electric current, either:

- **direct current, d.c.**, or
- **alternating current, a.c.**

You probably use both every day. When you use a mobile phone, laptop or mp3 player you will be using d.c. Direct current is usually produced by batteries or cells. The potential difference from the battery remains the same, and so the current through the appliance is in one direction only.

Electricity from the mains is described as a.c. With alternating current, the potential difference switches smoothly between positive and negative values as part of a repeating cycle. This causes the current to constantly change in size or reverse its direction very rapidly. The number of cycles per second is called the **frequency**. This is measured in hertz (Hz). In the UK the mains supply is 230 V a.c. with a frequency of 50 Hz, that is, 50 cycles per second.

The oscilloscope and a.c.

An **oscilloscope** may be used to produce an image showing how the potential difference of an electrical supply changes with time. With a d.c. supply the potential difference remains constant, and so the screen shows a horizontal line. When an a.c. supply is attached the potential difference changes, and so produces a trace which looks like a sine wave.

▲ Most electronic appliances need an adaptor to convert a.c. from the mains into the d.c. used by the device

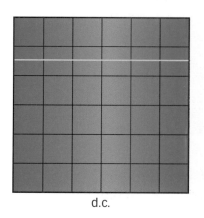
d.c.

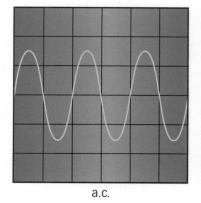
a.c.

▲ Traces on an oscilloscope from d.c. and a.c supplies look very different

---

A Give three examples of appliances that use a.c. and three that use d.c.

B What is the frequency of the UK mains supply?

The squares on the oscilloscope screen are a bit like the scale on a graph. Both the vertical and horizontal scales can be adjusted on the oscilloscope. The vertical squares measure the potential difference. If the scale is set to 5 volts per division (or 5 V/div) it would mean that each square represents 5 V. A 10 V d.c. supply would cause the green line to jump up two squares.

The horizontal scale is time. Each square can be set to be a number of seconds (or more commonly milliseconds). This is called the **time base**. A time base of 0.4 s/div would mean each square represents 0.4 s.

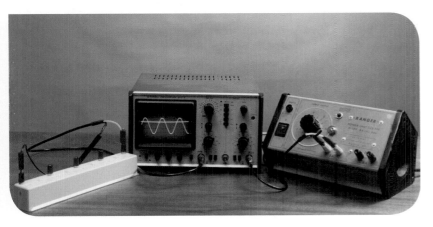

▲ An oscilloscope can be used to see how the potential difference of an a.c. supply changes with time

Finding the frequency of a supply

The frequency of an a.c. supply can be calculated if you know the time for one cycle. This is called the **time period**. Time period and frequency are related as follows in the equation below:

$$\frac{\text{frequency}}{\text{(hertz, Hz)}} = \frac{1}{\text{time period (seconds, s)}}$$

A time period of 2 s would correspond to a frequency of 0.5 Hz. A time period of 0.01 s would mean that the frequency is 100 Hz.

Using the trace on the screen of an oscilloscope, and knowing the time base, you can determine the time period of an a.c. supply. If the time base is set at 0.05 s/div and the cycle repeats every 2 squares, then the time period is 0.10 s (2 × 0.05 s). From the time period you can calculate the frequency, in this case 10 Hz.

Did you know...?

Mains voltage was not standardised in the UK until 1926. Until then different parts of the country had different potential differences and frequencies. If you bought an electrical device from one part of the country it might not have worked where you lived! The supply in the UK used to be 240 V a.c., but in 1993 it was changed to the current value of 230 V a.c. to match the rest of Europe.

Questions

1 What is the potential difference of the UK mains supply?

2 Sketch the trace that would be seen on an oscilloscope for:

(a) a d.c. battery

(b) an a.c. supply.

3 Calculate the frequency of the supply shown in the diagram if the time base is set at 0.02 s/div.

4 Sketch the trace seen on an oscilloscope screen if the time base is set to 0.01 s/div and the V/div is set at 200 V/div and it is connected to the UK mains supply.

Learning objectives

After studying this topic, you should be able to:

✔ understand the structure of the UK three-pin plug and its cables

✔ describe and compare some of the electrical safety features in the home (including circuit breakers and fuses)

✔ describe what happens when an appliance is earthed

Key words

earth, live, neutral, residual current circuit breaker, double-insulated

▲ A 3-core and a 2-core wire. Both contain a live and a neutral wire, but only the 3-core has an earth wire.

▲ Fuses usually come as replaceable cartridges. The wire inside the fuse heats up and melts if there is too much current.

The UK plug

If you've ever been on holiday abroad, you know that you had to take a plug adapter to use or recharge any of your electrical appliances. Different countries use different plug designs depending on their own electrical systems. The UK plug is unusual as it contain three connections (or pins).

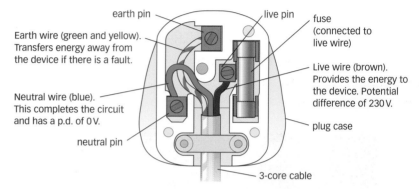

earth pin

Earth wire (green and yellow). Transfers energy away from the device if there is a fault.

Neutral wire (blue). This completes the circuit and has a p.d. of 0 V.

neutral pin

live pin

fuse (connected to live wire)

Live wire (brown). Provides the energy to the device. Potential difference of 230 V.

plug case

3-core cable

▲ A standard UK three-pin plug

3-core and 2-core cables

The cables connecting an appliance such as a TV to the mains are often described as 3-core wires as they contain three separate wires, each with a copper core. Copper is used as it is an excellent conductor. Some devices do not need an **earth** wire and these use 2-core cables. Every cable must contain both a **live** and **neutral** wire.

> **A** How many pins are there on a standard UK plug?
>
> **B** Why do most cables have a copper core?

Safety devices

There are several devices designed to improve the safety of the mains electricity supply.

Fuses

In the UK, each plug contains a fuse. This is connected to the live wire and is usually a small cylindrical cartridge.

Each fuse has a rating, and if the current passing through it exceeds this rating, the fuse heats up, melts, and breaks the circuit. This protects your electrical appliances from surges of current – the fuse melts before the wires in the computer.

A fuse with a rating of 13 A would melt if 13 A were to pass through it. A 3 A fuse contains a thinner wire which melts when the current reaches just 3 A. It is important to select the correct fuse rating for each appliance.

Earth wires

The earth wire is another important safety device. One end of this wire is connected to the metal case of the appliance. If the live wire were to come loose and touch the case, the case would become live. If you were then to touch the case, you could receive a very dangerous shock. With the earth wire attached, the case cannot become live as the current passes down the earth wire. This causes a surge in current, and this melts the fuse.

▲ Some appliances are double-insulated, as indicated by the square symbol in the photo. They do not need an earth wire as the case cannot become live.

Circuit breakers

Another safety device is a circuit breaker. Like the fuse, this is connected to the live wire of an appliance, but rather than heat up it detects tiny changes in current.

Residual current circuit breakers (RCCBs) detect a difference in the current between the live and neutral wire. If these values don't match, there must be a fault, and so the circuit breaker shuts off the current.

Despite being more complex than fuses, circuit breakers have a number of advantages. They switch the current off much faster than fuses, and they can be easily reset and used again.

Exam tip

✔ Don't just say a fuse 'blows' when there is too much current. You need to be more precise: it heats up, melts and breaks the circuit.

Did you know...?

Some appliances don't have an earth wire. These are usually made from non-conducting materials (like wood or plastic) and so the case cannot become live. Even some metal appliances are **double-insulated**. The live components are sealed away from the case and so there is no chance of the case becoming live.

Questions

1 Explain why plugs are made of rigid plastic.

2 List the three wires found in a UK plug, state their colour, and explain what they do.

3 Give two advantages of using circuit breakers compared with fuses.

4 Explain how the earth wire protects you if there is a fault with your device.

5 Describe how an RCCB protects a circuit.

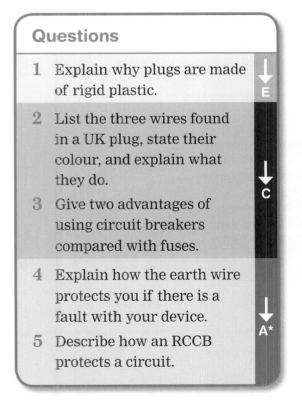

Learning objectives

After studying this topic, you should be able to:

- ✔ describe what happens when an electrical charge flows through a resistor, and relate this to suitable cable size
- ✔ calculate the power of an appliance
- ✔ use the equation linking energy transferred, potential difference, and charge

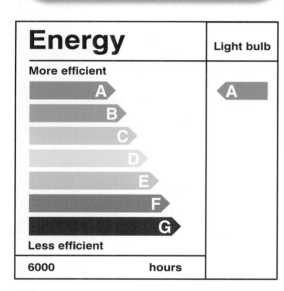

▲ Efficiency label for a light bulb

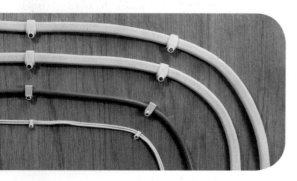

▲ Different sizes of power cable

> **A** A TV transfers 6600 J in 60 s. What is its power?

Resistance and efficiency

When an electrical charge flows through a resistor, the resistor gets hot. Energy has been transferred from the charge to the resistor.

Many electrical appliances waste energy by transferring electrical energy to heat energy. For example, filament lamps transfer a much higher percentage of electrical energy to heat energy than compact fluorescent lamps do.

When you are buying new appliances, you can check how much energy they waste by looking at the efficiency rating label on the packaging. An appliance which wastes less energy will have a higher rating. The most efficient appliances are rated A++.

Power cables

The size of a power cable for an appliance is related to the amount of energy it transfers. The cable needs to be large enough not to heat up when the current flows through it. An appliance that needs a higher power will have a larger cable.

A kettle with a power rating of 3 kW has a thick cable. The cables connecting a laptop to the mains have smaller thicknesses: the one leading to the mains socket supplies a power of 250 W, so it is thinner than the cable for the kettle, and the cable from the adapter to the laptop supplies only about 65 W, so it is thinner still.

Power

Different appliances transfer energy at different rates. The rate is given by the equation:

$$\frac{\text{power}}{\text{(watts, W)}} = \frac{\text{energy transferred (joules, J)}}{\text{time taken (seconds, s)}}$$

If power is P, the energy transferred is E, and the time taken for the energy transfer is t, then:

$$P = \frac{E}{t}$$

Power, potential difference, and current

The power of an appliance can also be calculated using the equation:

$$\frac{\text{power}}{\text{(watts, W)}} = \frac{\text{current}}{\text{(amperes, A)}} \times \frac{\text{potential difference}}{\text{(volts, V)}}$$

When power is P, current is I, and the potential difference provided by the supply is V, then:

$$P = I \times V$$

You can use this equation to calculate the current through an appliance and decide what size of fuse you should use for the appliance. The two most common sizes for fuses are 3 A and 13 A.

Worked example 1

A computer has a power rating of 200 W. What fuse should be fitted to the plug? Assume that the mains supply is 230 V.

power = current × potential difference or $P = I \times V$

$$\text{current} = \frac{\text{power}}{\text{potential difference}}$$

$$= \frac{200\ \text{W}}{230\ \text{V}} = 0.87\ \text{A}$$

The fuse with the smallest current rating that can be fitted to the plug is 3 A, so this should be used.

How much energy is transferred?

You might remember that the potential difference is the energy transferred (or work done) for each coulomb of charge passing through an appliance. So we can calculate how much energy is transferred.

$$\begin{array}{ccc} \text{energy} & & \text{potential} \\ \text{transferred} & = & \text{difference} \times \text{charge} \\ \text{(joules, J)} & & \text{(volts, V)} \quad \text{(coulombs, C)} \end{array}$$

If energy transferred is E, potential difference is V, and charge is Q, then:

$$E = V \times Q$$

Worked example 2

A charge of 50 C flows through a TV that is plugged into the mains. How much energy is transferred?

energy transferred = potential difference × charge

$$= 230\ \text{V} \times 50\ \text{C}$$

$$= 11\ 500\ \text{J}$$

B A toaster has a power rating of 1100 W. What fuse should be fitted to the plug?

C A 24 V battery transfers 75 C when starting an engine. How much energy is transferred to the engine?

Exam tip

✔ Remember that you should use watts rather than kilowatts in this type of calculation.

Questions

1 What are the two most common types of fuses? ↓ E

2 A food processor transforms 50 000 J of energy in 50 seconds. What is its power?

3 What size of fuse should be fitted to the following appliances (assume 230 V mains supply): ↓ C

　(a) a hair dryer with a power rating of 1500 W

　(b) a plasma TV with a power rating of 450 W?

4 (a) A charge of 5000 C flows through the hair dryer. How much energy is transferred?

　(b) A charge of 3000 C flows through the plasma TV. How much energy is transferred? ↓ A*

5 What links can you see between the equations for power, energy transfer, and charge?

Learning objectives

After studying this topic, you should be able to:

- ✔ describe the structure of an atom and the relative charges and masses of its components
- ✔ explain how the charge of an atom changes when it loses or gains electrons
- ✔ understand the meaning of the term isotope
- ✔ describe how new evidence caused scientists to change their model for the atom

	Proton	Neutron	Electron
relative mass	1	1	~0 (1/2000)
relative charge	+1	0	−1

A Which particle has a positive charge?

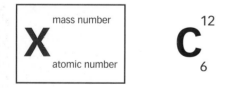

Each atom has a mass number and an atomic number

B Explain the meaning of the term isotope.

The atom

Atoms make up all matter. They are the smallest part of every substance. We now know that the atom has a very small central **nucleus**, with **electrons** in orbit.

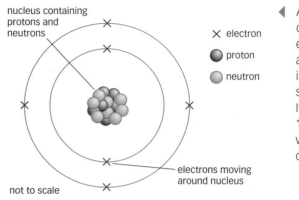

nucleus containing protons and neutrons

× electron
● proton
○ neutron

electrons moving around nucleus

not to scale

◄ All atoms have a central nucleus with electrons in orbit around it. The nucleus is around 10 000 times smaller than the atom. If the nucleus was 1 mm across, the atom would have a diameter of around 10 m!

Protons and **neutrons** are about the same size, but protons are positively charged and neutrons are neutral. Electrons are negatively charged, and much smaller than the particles in the nucleus.

All atoms have an **atomic number** and a **mass number**. The atomic number is the number of protons in the atom. All atoms of the same element have the same number of protons, and so the same atomic number. For example, all carbon atoms contain six protons.

The mass number refers to the number of protons plus the number of neutrons. A carbon 12 nucleus contains six protons and six neutrons (12 particles in total).

Isotopes and ions

Isotopes are atoms with the same number of protons, but a different number of neutrons. Isotopes of carbon include carbon 12 (six neutrons), carbon 13 (seven neutrons), and carbon 14 (eight neutrons). Each atom contain six protons – remember that all carbon atoms contain six protons.

Atoms are usually neutral. They have the same number of protons as electrons, so there is no overall charge. However, it is possible to add or remove electrons from atoms, creating charged **ions**. When ions have the same type of charge they repel each other, whereas oppositely charged ions attract.

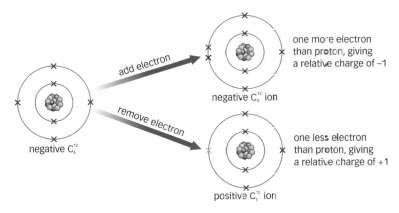

△ Adding or removing electrons creates ions

The discovery of the nucleus

Before the nucleus was discovered, scientists thought the atom was like a plum pudding. They thought the atom was a sphere with electrons scattered throughout it, rather like blueberries in a muffin. This changed when students of the famous physicist Ernest Rutherford collected some evidence from a scattering experiment.

They fired tiny positive particles, called alpha particles, at a very thin piece of gold foil (or gold leaf). They were extremely surprised to find that some of these particles were scattered back from the foil. Rutherford concluded that the atom must be very different from the 'plum pudding model'. Instead, he suggested that instead there must be a small, dense, positive centre. He called this the 'nucleus'. He knew that the nucleus must be small because most of the alpha particles went straight through the gold atoms. Scattering occurred only with the alpha particles that happened to be lined up with the nucleus and were repelled back the way they came.

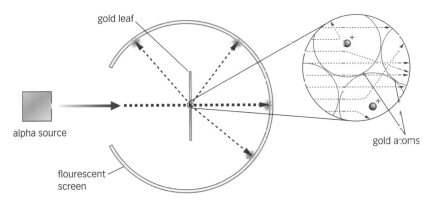

△ Although most alpha particles went straight through the gold leaf, some were scattered back

Did you know...?

Rutherford was right. The nucleus is really tiny; most of the atom is just empty space. The nucleus is so small that if it were the size of a pea and it was placed on the centre spot at Wembley stadium, the electrons would be orbiting in the surrounding car parks.

Questions

1 Draw a diagram of an atom, labelling all the important parts.

2 State the number of protons and the number of neutrons in an atom of U_{92}^{238}.

3 Explain what an ion is, and how one might be formed.

4 Describe how Rutherford discovered the nucleus.

5 Explain how Rutherford was able to conclude that the nucleus was small, dense, and positive.

E

C

A*

Learning objectives

After studying this topic, you should be able to:

- ✔ describe radioactive decay
- ✔ list some of the sources of background radiation
- ✔ understand the meaning of the term half-life, and how to determine half-life from a graph of activity against time

Key words

radioactive decay, ionising radiation, background radiation, activity, random, half-life

Radioactive decay and background radiation

Nuclear radiation is naturally around us all of the time; it is not only man-made. The nuclei of some atoms are unstable, and **radioactive decay** occurs when a nucleus from one of these atoms breaks down and emits one of the three types of **ionising radiation**:

- alpha particles (α)
- beta particles (β)
- gamma rays (γ).

The radiation around us is called **background radiation**. It comes from a variety of sources. Less than 15% comes from man-made activities:

- use of radiation in hospitals
- nuclear weapons testing
- nuclear power, including accidents.

Most of it occurs naturally due to the breakdown of radioactive atoms found within rocks and from cosmic rays arriving from space.

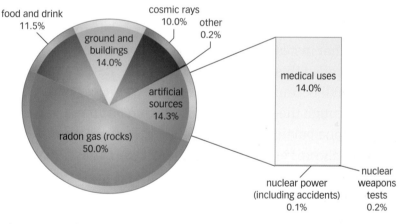

▲ Sources of background radiation found in the UK

A What are the three most common sources of background radiation in the UK?

Your exposure to radiation depends on where you live and, for example, on the nature of your job. Different parts of the country have different levels of background radiation, and an airline pilot or someone working in a hospital can receive a higher dose than other people.

Detecting radiation

Radiation may be detected using a Geiger–Muller (GM) tube. When the radiation enters the tube, it ionises the atoms of the gas inside the tube. It removes electrons from the atoms in the gas and this causes a small electric current which is detected by a counter.

If atoms within human cells are ionised this can damage the cell. The DNA within the cell can be affected. This might ultimately kill the cell or cause it to mutate, potentially leading to cancer.

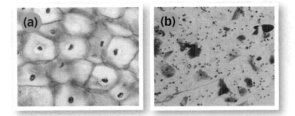

▲ (a) Healthy cells, and (b) cells that have been exposed to a high dose of ionising radiation

> **B** What effect does ionising radiation have on human cells?

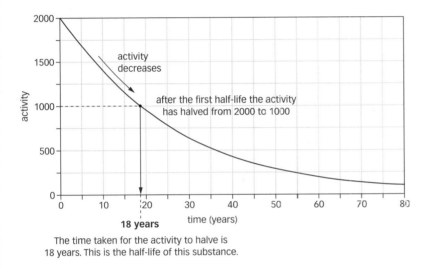

The time taken for the activity to halve is 18 years. This is the half-life of this substance.

▲ Half-life can be determined from a graph of activity against time

Half-life

A substance which contains radioactive atoms gives out radiation all of the time. The nuclei within it decay, and the number of these decays per second is called the **activity**. You cannot stop a radioactive substance decaying, no matter what you do to it. This radioactive decay is also a truly **random** process. There is no way of knowing which nuclei will decay next, nor when any particular atom will decay.

As the radioactive atoms within a substance decay, fewer and fewer radioactive nuclei remain. This leads to a drop in activity as time passes. The average time taken for half of the radioactive nuclei in a sample to decay is called the **half-life**. This is the time it takes for the activity to halve.

Different isotopes have different half-lives. These range from milliseconds to billions of years. The table below shows a few examples.

Questions

1 Name the three types of ionising radiation.

2 Explain what is meant by the term ionising radiation.

3 Explain what is meant by half-life and describe how you can find the half-life from a graph of activity against time.

4 What fraction of radioactive nuclei remain in a sample after:
   (a) one half-life?
   (b) two half-lives?
   (c) six half-lives?

5 What proportion of carbon 14 will be left in a sample after 28 500 years?

Isotope	Half-life
nitrogen 17	4 seconds
radon 220	3.8 days
carbon 14	5700 years
uranium 238	4.5 billion years

Learning objectives

After studying this topic, you should be able to:

✔ describe some properties of alpha, beta, and gamma radiation in terms of the changes in the nucleus and their penetrating power

✔ describe how to handle radioactive sources safely

Key words

alpha particle, beta particle, gamma ray

A State the three types of radiation and give examples of materials that stop them.

Did you know...?

In November 2006 Alexander Litvinenko, a former officer in the Russian Federal Security Service, died in University College Hospital London. It is believed he was murdered. The cause of death was radiation poisoning from polonium 210. He consumed just 10 millionths of a gram. It was not possible to detect the radiation as the alpha particles did not leave his body. However, they caused fatal damage to his internal organs.

Types of decay and penetrating power

The three types of ionising radiation all come from the nucleus of unstable atoms, but they are each very different.

alpha particles α_2^4	These are very ionising particles made up of two protons and two neutrons. This is the same as a helium nucleus. Alpha particles have a large mass and a large positive charge. This makes them the most ionising type of radiation.
beta particles β_1^0	A beta particle is a fast electron from the nucleus. Beta particles have a negative charge. They have less mass and a lower charge than alpha particles, making them less ionising.
gamma rays γ_0^0	A gamma ray is a high frequency electromagnetic wave. This kind of radiation is not very ionising, but travels a long way.

How far the radiation travels depends on how ionising it is. The more ionising it is, the more quickly it slows down, and stops. Alpha radiation travels only a few centimetres through air, beta particles a few metres and gamma radiation several kilometres.

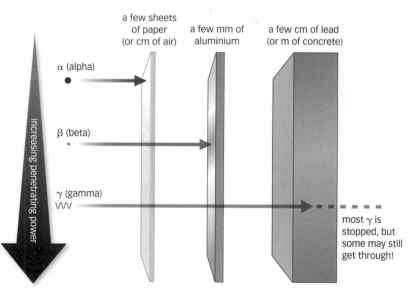

▲ The different penetrating powers of the three types of radiation

Deflection of radiation

Alpha and beta radiation are both deflected by electric and by magnetic fields. Alpha particles are deflected less than beta particles, and they are deflected in the opposite direction from beta particles. Gamma radiation is not deflected.

This is explained by the charge and mass of the radiation. Alpha and beta radiation have opposite charges, so they are deflected in opposite directions. Gamma radiation has no charge, so it is not deflected. Alpha particles have a greater mass than beta particles, so alpha radiation is deflected less.

Nuclear equations

When a radioactive nucleus emits an alpha particle, the mass number and atomic number of the nucleus changes. The mass number drops by 4 and the atomic number drops by 2. This means that the element changes. For example, uranium 238 becomes thorium 234 after emitting an alpha particle.

$$U_{92}^{238} \rightarrow Th_{90}^{234} + \alpha_{2}^{4}$$

In beta decay a neutron breaks up into a proton and a beta particle. As a result the mass number of the atom stays the same (it has lost one neutron but gained one proton), but the atomic number goes up by 1. For example, carbon 14 forms nitrogen 14 after beta decay.

$$C_{6}^{14} \rightarrow N_{7}^{14} + \beta_{-1}^{0}$$

These decay equations must always balance, with the atomic numbers and mass numbers adding up to the same value on both sides.

Handling radioactive sources safely

Radioactive sources must be handled very carefully. Depending on the type of radiation and the activity of the radioactive source, different precautions are needed. In each case you need to think about the shielding, the length of time of exposure and the distance from the source.

Shielding can be as simple as wearing gloves, but if the source emits more penetrating radiation then denser shielding such as lead, lead crystal glass, or special radiation suits might be needed.

To reduce the time you are exposed to radiation, sources are only taken out when they are being used.

You should always keep your distance from radiation sources. The further away you are from the source the lower your exposure. Using tongs keeps them at a safer distance.

▲ An example of the clothing worn to protect against strong radioactive sources

Questions

1 List the three types of radiation in order of their penetrating power, from most penetrating to least penetrating. ↓ E

2 Describe some of the safety precautions needed when handling radioactive sources. ↓ C

3 Using a GM (Geiger–Müller) tube and an assortment of different absorbers, describe an experiment you could do to determine the types of radiation emitted from a radioactive source. ↓ A*

4 Complete the following decay equations:
   (a) $Pu_{94}^{239} \rightarrow U_{?}^{?} + \alpha_{?}^{?}$
   (b) $Pb_{82}^{210} \rightarrow Bi_{?}^{?} + \beta_{?}^{?}$

Learning objectives

After studying this topic, you should be able to:

✔ describe some of the uses of radiation

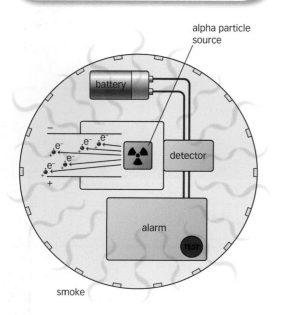

▲ When smoke absorbs the particles, the alarm goes off

Key words

gamma knife, sterilising

B Describe what would happen to the recorded counts if the paper was too thin.

Smoke alarms and alpha particles

Alpha particles don't travel very far, but are very ionising. They are ideal for use in smoke alarms. Most smoke alarms contain a weak source of alpha radiation. This ionises air inside the alarm and creates a very small electric current (just like in the GM tube). When smoke enters the alarm this current drops, setting off the alarm. If there is no smoke, there is no change in current, and so the alarm stays quiet.

The source in the alarm must have a long half-life – the activity must remain fairly constant, so that you will not need to replace it regularly.

A Describe how radiation is used to detect smoke in a smoke alarm and why alpha sources are used.

Beta particles and paper mills

Compared with alpha radiation, beta particles can pass through thicker materials. Most beta particles are able to travel through several sheets of paper. In paper mills, the sheets of paper pass between a beta source and a detector.

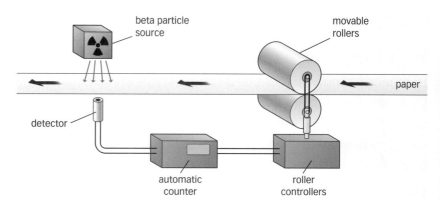

▲ Beta radiation may be used in paper mills to automatically monitor the thickness of the paper

The thicker the paper, the fewer beta particles get through. This makes the counts recorded by the counter go down. If the counts fall too low, this means that the paper is too thick and so the rollers squeeze together.

As with the smoke alarm, the source selected must have a long half-life to keep the activity fairly constant.

Medical and other uses of gamma rays

An important use of gamma rays is in killing cancerous cells within the body as part of radiotherapy. Doctors may use a special machine called a **gamma knife.** This contains a movable source of gamma rays that are focussed on the tumour and are fired into the body. The gamma source is moved around in order to reduce the exposure of the healthy tissue to the radiation while still providing a high enough dose to kill the cells inside the tumour.

Gamma radiation is used for **sterilising** medical equipment. It kills microorganisms such as bacteria

Gamma rays are also used in medical tracers. A weak source of gamma rays is either ingested by the patient or injected. Gamma rays travel out of the body and special cameras are used to monitor the movement of the source around the body. Doctors can then identify problems such as blockages or leaks within internal organs.

A similar technique can also be used to detect leaks in underground pipes.

In both cases, gamma sources with fairly short half-lives are used, usually just a few hours. This means the radioactivity falls to a low level very quickly.

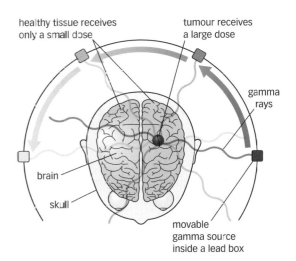

healthy tissue receives only a small dose

tumour receives a large dose

gamma rays

brain

skull

movable gamma source inside a lead box

▲ A gamma knife is used to kill cancerous cells in a brain tumour

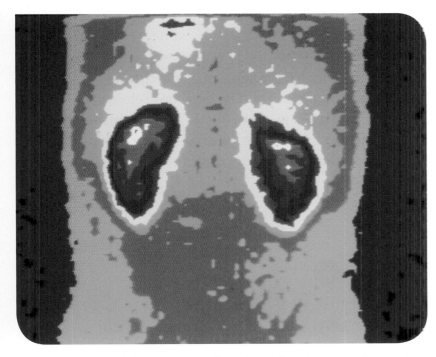

▲ An image of the kidneys obtained by using radioactive tracers injected into the body

Questions

1 Give three different examples of uses of radiation. ↓ E

2 A similar arrangement to the one used in the paper mill is used to ensure steel sheets are kept at a constant thickness. Explain how this process might work, describing the type of radiation you would use. ↓ C

3 Give two reasons why gamma sources are used as medical tracers rather than alpha or beta sources, and explain why the source must have a short half-life.

4 Explain why a beta source with a long half-life is used in paper mills. What would happen if a source with a short half-life were to be used instead? ↓ A*

5 Describe the advantages and disadvantages of using a radioactive medical tracer to diagnose a serious medical condition.

Learning objectives

After studying this topic, you should be able to:

✔ describe the process of nuclear fission

✔ describe the process of nuclear fusion

✔ compare the advantages and disadvantages of each method for producing energy

Key words

nuclear fission, uranium 235, plutonium 239, chain reaction, nuclear fusion

A Give two examples of fissionable materials.

B Describe how a nuclei of uranium 235 may be split and give the name of this process.

Did you know...?

In a nuclear reactor the chain reaction is carefully controlled. In a nuclear bomb, the number of fissions increases dramatically, releasing vast amounts of energy in a very short time. The Russian Tsar bomb released the energy equivalent of exploding 50 million tonnes of TNT. That is an incredible 2.1×10^{17} J, enough to power over 55 million TVs for a year.

Nuclear fission

Nuclear fission is the splitting of atoms. It is the reaction which takes place inside all nuclear reactors. Fission releases energy in the form of heat. This is used to turn water into steam which turns turbines, which then turn generators to produce electricity.

Either **uranium 235** or **plutonium 239** atoms are split into two smaller nuclei. Most reactors use uranium 235. These smaller nuclei are often very radioactive, and make up the radioactive waste produced by nuclear reactors.

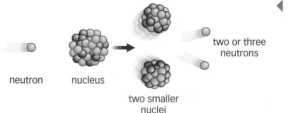

neutron   nucleus   two smaller nuclei   two or three neutrons

◄ When a fissionable nucleus absorbs a neutron, it splits into two smaller nuclei, and fires out two or three neutrons in its turn

In nuclear fission, a nucleus first absorbs an extra neutron. This makes the nucleus spin and distort. After a few billionths of a second it splits into two smaller nuclei and releases two or three further neutrons.

Uranium 235 and plutonium 239 are described as fissionable substances as they can both be split easily.

Chain reactions

If there are enough fissionable nuclei in the material, then a **chain reaction** may start. The neutrons released in the first fission are absorbed by more nuclei, which then also split. These fissions release more neutrons, which lead to even more fissions and the process continues.

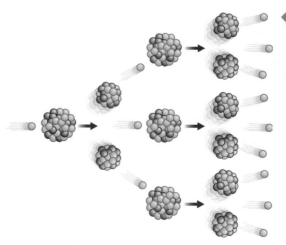

◄ In a chain reaction the neutrons from one fission go on to create further fissions

Nuclear fusion

In **nuclear fusion** atomic nuclei are stuck (or fused) together. This forms heavier nuclei. Like nuclear fission, this process releases heat energy. Nuclear fusion is the process by which energy is released in all stars, including our Sun.

Use of nuclear fusion to generate electricity in the future would have the same advantages as fission nuclear power. No carbon dioxide is produced and very large amounts of electricity can be generated. Unlike nuclear fission, nuclear fusion does not produce radioactive waste.

Fusion on Earth

So far it has proved to be very difficult to sustain a nuclear fusion reaction on Earth. Atomic nuclei are positively charged because of their protons. This means they repel one another other when they get close together. The nuclei have to move very, very fast to get close enough to fuse. This happens in the core of stars like our Sun because the core is so hot. The nuclei are moving around at very high speeds and smash together.

Several experimental fusion reactors are being built. Some use superstrong magnetic fields to try to squeeze the nuclei together. Others use incredibly powerful lasers to heat up a tiny volume of gas to enormous temperatures.

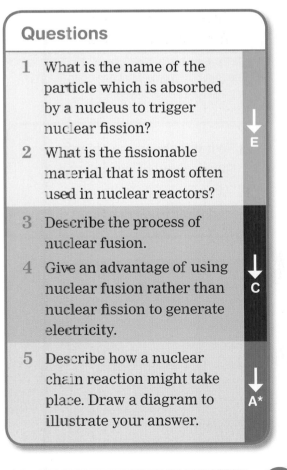

two small nuclei

one larger nucleus

▲ In nuclear fusion two smaller nuclei fuse together to make a larger nucleus

▲ The interior of the experimental JET Tokamak fusion reactor in Oxfordshire

Exam tip **AQA**

✔ Do not confuse fission and fusion with radioactive decay (alpha, beta, or gamma). Nuclear reactors or nuclear bombs are not an example of a use of radioactivity.

✔ Be careful not to mix up fission and fusion. In nuclear *fusion* nuclei are *fused* together.

Questions

1 What is the name of the particle which is absorbed by a nucleus to trigger nuclear fission? ↓ E

2 What is the fissionable material that is most often used in nuclear reactors?

3 Describe the process of nuclear fusion.

4 Give an advantage of using nuclear fusion rather than nuclear fission to generate electricity. ↓ C

5 Describe how a nuclear chain reaction might take place. Draw a diagram to illustrate your answer. ↓ A*

Learning objectives

After studying this topic, you should be able to:

- ✔ explain how stars are formed
- ✔ describe what happens when a star approaches the end of its life
- ✔ describe the complete life cycle of a star

▲ The Sun is in the main phase of its life. This will last for billions of years.

A What is the name given to a large cloud of gas in space?

B What force pulls the gas together to form a star?

Did you know...?

It is amazing to think you are made of star dust. All the elements in your body above hydrogen were made in the core of large stars. The atoms of elements heavier than iron were created when these stars went supernova, blowing these atoms into space to eventually form you!

Star birth and life

The Universe contains billions of stars. All of them are formed in the same way. A huge cloud of gas (mainly hydrogen) called a **nebula** begins to be pulled together by gravity. A large ball of gas forms in the centre of this cloud.

As it gets denser, more gas is pulled in. The ball gets hotter and hotter, forming a **protostar** (a similar process forms planets around the star). Eventually the protostar gets so big, and its centre gets so hot, that nuclear fusion happens in its core. A star is now born. Hydrogen nuclei are fused together to make helium.

The energy released in nuclear fusion pushes out against gravity and keeps the star stable. These forces are balanced. This stage is described as the main sequence of the star, and it can last for billions of years. Our Sun has been in its main sequence for around 5 billion years. It has another 5 billion years to go before it runs low on hydrogen in its core, and begins to die.

Dying stars

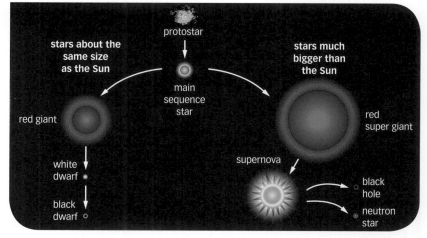

▲ The life cycle of stars

Smaller stars

All stars eventually die, but what happens to them depends on their mass. Small stars, such as our Sun, gradually cool and expand into a **red giant**. The outer layers of the star break away and form a planetary nebula. All that remains is the white hot core of the star. This **white dwarf** gradually cools over time.

Larger stars and new elements

Stars that are much larger than our Sun have much more dramatic lives. They too eventually expand and cool, but they grow much bigger, turning into **red super giants**. They are so big that they are able to fuse heavier nuclei together to make other elements. When the Universe was formed in the Big Bang it was mainly made of hydrogen – the heavier elements, up to iron, have all been made in the core of dying stars.

The star then explodes in a gigantic explosion called a **supernova**. When a star goes supernova it often outshines all the other stars in the galaxy put together. All the elements heavier than iron are created here. This gigantic explosion spreads these elements throughout the Universe.

Neutron stars and black holes

During a supernova, the core of the star is crushed down by large gravitational forces. This can form a very dense kind of star that is made up only of neutrons. This **neutron star** spins very fast and sends pulses of radio waves through space that can be detected on Earth.

If the star is even bigger, the core is crushed down into a tiny space – it forms a **black hole**. These mysterious objects have infinite densities. All the mass of the core is crushed down into a space smaller than an atom. The gravitational force is so large that nothing can escape, not even light.

▲ The Crab Nebula is left over from a star that went supernova in 1054

C What is the name given to the explosion of a large star at the end of its life?

Key words

nebula, protostar, red giant, white dwarf, red super giant, supernova, neutron star, black hole

Questions

1 What is the name given to the hot remains of a red giant star after the outer layers have broken away, leaving just the core?

2 Describe why our Sun will remain stable for billions of years.

3 Explain how a large star first produces heavier elements, and then spreads them throughout the Universe.

4 Describe the life cycle of a star:
   (a) The size of our Sun
   (b) A star much bigger than our Sun.

↓E

↓C

↓A*

Exam tip

✓ What happens to a star depends on how big it is (its mass). Our Sun is too small to supernova and form a black hole. Only stars with much more mass explode at the end of their lives.

Course catch-up

Revision checklist

- Some materials gain an electrostatic charge when rubbed together. Opposite charges attract, like repel.
- Electric current is the flow of charge around a circuit.
- The potential difference between two points in an electronic circuit is the work done (energy transferred) per coulomb of charge that passes through the points.
- Circuit symbols represent electrical components.
- Resistance is a measure of how difficult it is for electrons to pass through a component.
- Current must pass through all components in a series circuit one after the other. Current in a parallel circuit can take more than one route.
- The resistance of a filament lamp increases as its temperature increases.
- An LED (light-emitting diode) produces light when a current flows through it, and is efficient and long-lasting.
- The resistance of an LDR (light-dependent resistor) varies with intensity of light. LDRs are used in switches.
- The resistance of a thermistor varies as its temperature changes. Thermistors are used in temperature sensors.
- Electric current has two forms: direct current (d.c.), usually produced by batteries or cells; and alternating current (a.c.), such as UK mains electricity (230 V a.c. with frequency 50 Hz).
- UK three-pin plugs contain a live wire (brown), a neutral wire (blue), an earth wire (green and yellow), and a fuse. Fuses, earth wires, and circuit breakers are safety devices.
- When an electrical charge flows through a resistor, energy is transferred. Larger appliances need larger power cables. The power of an appliance is its rate of energy transfer.
- Atoms have a small central nucleus of protons (positively charged) and neutrons (no charge) surrounded by negatively charged electrons in orbit. Positively or negatively charged ions are created when atoms gain or lose electrons.
- Isotopes are atoms with the same number of protons (same atomic number) but a different number of neutrons.
- Radioactive decay is the breakdown of the nucleus of an atom to form ionising radiation. Half-life is the average time taken for activity (decays per second) to halve.
- Nuclear fission is used in nuclear reactors to produce electricity.
- Nuclear fusion produces energy in stars.

power generation

nuclea fission

stars

nuclear fusion

power = current × potential difference

power, P (watts, W)

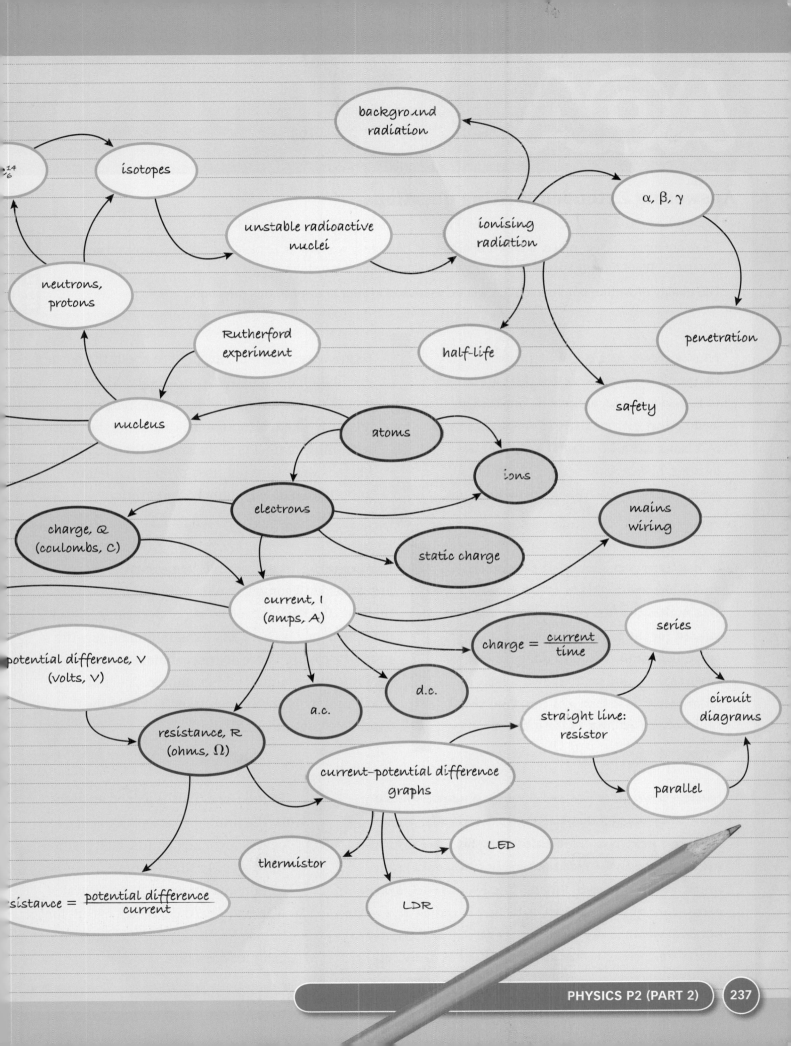

background radiation

isotopes

$^{14}_{6}$

unstable radioactive nuclei

ionising radiation

α, β, γ

neutrons, protons

Rutherford experiment

half-life

penetration

nucleus

atoms

ions

safety

charge, Q (coulombs, C)

electrons

static charge

mains wiring

current, I (amps, A)

charge = $\dfrac{current}{time}$

series

potential difference, V (volts, V)

a.c.

d.c.

circuit diagrams

resistance, R (ohms, Ω)

current-potential difference graphs

straight line: resistor

parallel

thermistor

LED

resistance = $\dfrac{potential\ difference}{current}$

LDR

Answering Extended Writing questions

QUESTION

There are many nuclear power stations operating successfully worldwide. Explain the process involved and discuss some of the arguments for and against building more such stations.

The quality of written communication will be assessed in your answer to this question.

G–E

Nuclear power gets electricity from radioactivity from uranum there are lots of α, β, and γ flying around hiting things and geting hot. The stuff is dangerous becos if it leaks out an you eat it you get sick and die. I don't want it near me! But its good becose it doesn't make grenehous gas.

Examiner: The facts given here about the power production are mostly wrong, though there is mention of uranium and heat. One reason in favour is given, although lacking detail; but the reasoning against is more tabloid than scientific. Physics words are not used properly. Spelling, punctuation, and grammar are erratic.

D–C

In nuclear power, atoms hit each other and brake apart. This is called fision. This makes the fuel get very hot, and that is used for the power. The electricity is chepe and clean. But It is dangerous because they can explode, and poison a lot of people, also terorists can steal the radioactive fuel and make a bomb.

Examiner: This answer has some correct ideas, and deals with some key points. But detail is missing: the fission process is not accurately described; nor is there any explanation of why the power is 'clean'. There are occasional errors in spelling, punctuation, and grammar. Crucially, the word 'fission' is ambiguously spelt – could the candidate have meant 'fusion'?

B–A*

Nuclear power uses fission. A moving neutron hits a uranium nucleus; this splits into two smaller atoms, and more neutrons are released. They hit other uranium nuclei, and a chain reaction happens. It causes heat, which makes the power. It doesn't use fossil fuel, and no greenhouse gas is produced. But there is radioactive waste material which is difficult to store safely; and there is a risk of dangerous material leaking like at Chernobyl.

Examiner: This is a good answer. In the limited space available it covers most key points, addressing the process and also some arguments for and against. Physics words are used correctly, except for 'atom'. Spelling, punctuation, and grammar are fine.

Exam-style questions

1 Match these quantities with their units.

A01

current	watts
potential difference	ohms
charge	joules
resistance	coulombs
power	amps
energy	volts

2 A circuit is shown below.

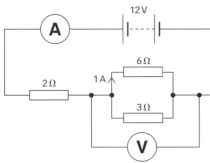

A01 **a** How are the 3Ω and 6Ω resistors connected together?

A01 **b** How are the Ω resistor, the ammeter, and the battery connected together?

A02 **c** Calculate:
   i the reading on the voltmeter
   ii the current through the 3 W resistor
   iii the current through the ammeter.

3 A cathode-ray oscilloscope (CRO) trace is shown below.

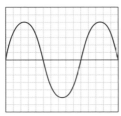

A01 **a** Explain why this trace illustrates an alternating current.

A01 **b** Describe how a d.c. trace would appear.

A02 **c** The time base of the CRO is set at 0.01 ms/division. Calculate:
   i the time period of the a.c.
   ii the frequency of the a.c.

A02 **d** If the frequency of the trace was doubled, how would the trace change?

Extended Writing

4 Explain how fuses, circuit breakers,
A01 and the earth wire help to keep mains electricity safe.

5 You have three radioactive isotope
A02 sources available, emitting α, β, and γ radiations. Explain which you would use in a smoke detector, a paper-thickness detector, and as a tracer to find where an underground pipe is leaking.

6 Describe the life cycle of a star about
A01 the same size as our Sun, and a star much bigger than our Sun.

A01 Recall the science
A02 Apply your knowledge
A03 Evaluate and analyse the evidence

Glossary

accelerate To speed up, slow down, or change direction.

acceleration Speeding up, slowing down, or changing direction. Change in velocity per second, measured in m/s^2.

activation energy Minimum amount of energy that particles need in order to react when they collide.

activity Number of radioactive decays per second.

actual yield Amount of product found by experiment.

aerobic Using/in the presence of oxygen.

air resistance Frictional force that slows down objects moving through air.

alkali Soluble base. Soluble hydroxides neutralise acids.

alkali metal An element in Group 1 of the periodic table (lithium, sodium, potassium, rubidium, caesium, francium).

allele Version of a gene.

alloy Mixture of a metal with one or more other elements. The physical properties of an alloy are different from those of the elements in it.

alpha particle Very ionising, but not very penetrating form of ionising radiation. Made up of 2 protons and 2 neutrons (a helium nucleus).

alternating current, a.c. Current that continually reverses its direction.

amino acids Molecules containing carbon, hydrogen, nitrogen, and oxygen. Many amino acids join together to make a protein.

amylase Enzyme that catalyses the breakdown of starch to sugar molecules.

anaerobic Without using/not in the presence of oxygen.

anhydrous Describes a salt that does not have its water of crystallisation.

antibodies Special proteins in the body that can bind to a particular antigen and destroy a particular pathogen.

asexual reproduction Reproduction without gametes/sex cells, using mitosis.

atomic number The number of protons in the nucleus of an atom.

attract Pull towards.

background radiation Radiation around us all of the time from a variety of natural and man-made sources.

bacteria Single-celled microorganisms, 1–5 μm in length. The DNA is not enclosed in a nucleus. Bacterial cells have cytoplasm, a cell membrane, and a cell wall.

base Metal hydroxide or oxide. Bases neutralise acids.

battery Scientific name for two or more cells connected together.

bauxite The most common ore of aluminium, containing the mineral aluminium oxide.

beta particle Ionising form of radiation. It is a fast electron from the nucleus.

black hole A very small, dense object with gravity so strong not even light can escape. The result of the supernova of a massive star.

braking distance Distance moved by a vehicle when it slows down and stops after the brakes have been applied.

braking force Used to stop or slow down a moving object such as a car.

breathing Movements of rib cage and diaphragm that cause air to enter and leave the lungs.

brine Concentrated sodium chloride solution.

buckminsterfullerene Form of carbon consisting of molecules containing 60 carbon atoms joined together to form a hollow sphere.

carbohydrase Enzyme that catalyses the breakdown of large carbohydrate molecules such as starch to smaller sugar molecules.

catalyse To speed up a reaction using a catalyst.

catalyst Substance that speeds up a reaction without being used up in the reaction.

cell (biological) Building block of living things.

cell (electrical) A source of electricity made from two plates of different metals separated by a conducting solution.

cell membrane A thin layer around a cell that controls the movement of substances into and out of the cell.

cell wall Rigid cellulose layer outside the cell membrane in a plant or bacterial cell.

chain reaction One reaction going on to create another, which creates another, and so on, such as a nuclear fission chain reaction inside a nuclear reactor.

charge Amount of electricity.

chlorophyll Green substance found in chloroplasts, where light energy is trapped for photosynthesis.

chloroplast Small disc in the cytoplasm of plant cells, containing chlorophyll. Photosynthesis occurs in chloroplasts.

chromatogram The record obtained in chromatography.

chromosome Structure in a cell nucleus that consists of one molecule of DNA that has condensed and coiled into a linear structure.

collide Particles collide when they come into contact as one or both of them is moving.

community All the populations of organisms that live together and interact in the same area.

compound Substance made up of two or more different elements, chemically combined.

concentration gradient Difference in concentration of a substance from one region to another.

conventional current The flow of positive charge in a circuit from high voltage to lower voltage, in the opposite direction to the actual flow of electrons (or electron current).

coulomb Unit of electric charge.

covalent bond Shared pair of electrons that holds two atoms together.

cryolite Substance in which aluminium oxide dissolves in an aluminium electrolysis cell, so that the process can take place at a lower temperature than would be the case if pure aluminium oxide were used.

crystallisation Process by which solid crystals are obtained from solution.

current Movement of charged particles (usually electrons) through a material.

current-potential difference graph Graph with current on the y-axis and potential difference on the x-axis. It shows how the current varies for different potential differences.

cytoplasm Jelly-like contents of a cell, inside the cell membrane. Cellular structures are suspended in the cytoplasm, and this is where many of the cell's reactions occur.

deceleration Negative acceleration, acceleration in the opposite direction to the direction of motion, so slowing down.

delocalised Describes electrons that are free to move throughout a structure.

denatured Describes the state of a protein when its shape has altered and it can no longer carry out its function.

detergent Substance added to water to improve its cleaning properties.

diamond Form of the element carbon in which each carbon atom is joined to four other carbon atoms by strong covalent bonds in a giant covalent structure.

differentiation Development of cells into types that are specialised for a particular function.

diffusion The spreading of the particles of a gas or a substance in solution, resulting in a net movement from a region of high concentration to a region of lower concentration. The bigger the difference in concentration, the faster the diffusion happens.

digestion Breaking down of large molecules into smaller, soluble particles that can be absorbed.

digestive system Organ system that digests food into smaller particles and absorbs these into the body.

digit Finger or toe.

diode Circuit component that conducts electricity in one direction only. It has a very high resistance in the reverse direction.

diploid Describes a cell that has a nucleus with two sets of chromosomes; a body cell.

direct current, d.c. Current that goes in only one direction.

directly proportional Describes the relationship between two quantities when one is always equal to the other but multiplied by a constant number.

displayed formula Description of a covalently bonded compound or element that uses symbols for atoms and that also shows the covalent bonds between the atoms.

distance–time graph Graph showing distance on the y-axis and time on the x-axis. It shows the distance travelled from a certain point at a particular moment.

distribution Detail of where species are found over the total area where they occur. For example, woodlice may have a high distribution under a log.

DNA (deoxyribonucleic acid) Chemical that carries the genetic code.

DNA bases Molecules arranged in pairs within each molecule of DNA. A pairs with T and C pairs with G.

DNA fingerprinting Technique that analyses parts of the DNA of an individual and compares it with that of other individuals/DNA samples to find out whether someone committed a crime, or to establish whether individuals are related.

dominant Visible characteristic present in an organism even when only one allele of the gene is present.

dot and cross diagram Diagram that shows how the electrons are arranged in a molecule or in ions.

double covalent bond Strong bond between two atoms that consists of two shared pairs of electrons.

double-insulated Describes an appliance in which all the live components are sealed away from the case, so the case cannot become live.

earth Pin/wire that carries energy safely away from the device to the ground if there is a fault.

elastic potential energy Energy stored in stretched or compressed objects.

electrode Conductor through which an electric current enters or leaves a melted or dissolved ionic compound in electrolysis.

electrolysis Process by which melted or dissolved ionic compounds are broken down by passing an electric current through them.

electrolyte Solution of an ionic substance that is broken down in an electrolysis cell.

electron Sub-atomic particle found outside the nucleus of an atom. It has a charge of −1 and a very small mass compared to protons and neutrons.

electroplating Covering an object with a thin coating of metal during an electrolysis reaction. The object to be plated is the negative electrode in an electrolysis cell.

electrostatic charge Charge from electrons that have been moved to or from an insulator.

endangered Describes a species that has low numbers and is in danger of becoming extinct.

endothermic reaction Reaction in which energy is transferred from the surroundings to a reacting mixture.

enzyme Biological catalyst made of protein. Enzymes catalyse chemical reactions in living organisms.

epidermal tissue Tissue one cell thick on the surface of plant roots, stems, and leaves, that protects the organs.

equal and opposite Describes the pair of forces produced when objects interact: equal in size and acting in opposite directions.

evolution Gradual change in an organism over time.

exothermic reaction Reaction in which energy is transferred to the surroundings, in the form of heat, light, and sound, for example.

extension Change in length of an object when a force is applied.

extinction End of a species, when all its members have died out.

F1 generation First filial (daughter) generation: the offspring from a genetic cross between two true-breeding parents.

fatigue Build-up of lactic acid in muscles that stops them contracting efficiently.

fermenter Large container used for growing large numbers of microorganisms.

fertilisation Fusion of male and female gamete nuclei.

filament lamp Lamp containing a coil of wire that glows when an electric current flows through it.

fluid Liquid or gas.

force Push or pull that changes the way an object is moving, or its shape. Measured in newtons.

formula Description of a compound or an element that uses symbols for atoms. It shows how many atoms of each type are in the substance.

fossil Preserved remains of ancient living things.

frequency Number of cycles of potential difference per second for an alternating current.

friction Force that acts to stop or slow down two objects that are sliding against each other.

fructose Type of sugar, which tastes sweeter than glucose.

fullerene Form of the element carbon based on hexagonal rings of carbon atoms.

fungi Organisms with cells containing a membrane, cytoplasm, nucleus, and a cell wall. The fungal cell wall is made of chitin rather than the cellulose of a plant cell wall.

fuse Thin piece of wire that melts (and breaks the circuit) if too much current flows through it.

gametes Sex cells – eggs and sperms.

gamma knife A special machine found in hospitals that uses gamma rays to kill cancerous tumours by focussing them on the tumour and minimising the exposure to the surrounding tissues.

gamma ray Very penetrating but not very ionising form of ionising radiation. It is a high frequency electromagnetic wave.

gas chromatography Method of separating and identifying a mixture of substances, and comparing the amounts of the different substances in the mixture. Helium or nitrogen gas carries chemicals from a sample through a chromatography column.

gene Length of DNA that codes for a characteristic/protein.

giant covalent structure Three-dimensional network of atoms that are joined together by covalent bonds in a regular pattern.

giant ionic lattice Regular, three-dimensional pattern of ions held together by strong electrostatic forces.

giant metallic structure Regular, three-dimensional pattern of positively charged metal ions held together by a sea of delocalised electrons.

glucose Type of sugar.

glycogen Large insoluble carbohydrate molecule; similar to starch but found only in animal cells, some bacterial cells, and some fungi.

gradient Slope of a graph.

graphite Form of the element carbon in which each carbon atom is joined to three other carbon atoms by strong covalent bonds to form a layer of carbon atoms. There are many of these layers in a piece of graphite. The layers are held together by delocalised electrons that move between the layers.

gravitational field strength Strength of the force of gravity on a planet. On Earth it is 10 N/kg.

gravitational potential energy Energy an object has when it is above the ground.

greenhouse Structure where plants can be grown under controlled conditions.

growth Increase in size, usually with an increase in cell numbers.

half equation Equation that describes the loss or gain of electrons by a reactant atom or ion, but is not balanced in terms of electrical charge.

half-life The time taken for half of the radioactive nuclei in a substance to decay, or the time taken for the activity from a substance to halve.

halogen An element in Group 7 of the periodic table (fluorine, chlorine, bromine, iodine, astatine).

haploid Describes a cell that has a nucleus with only one set of chromosomes; a sex cell.

heart rate Number of times per minute a heart beats.

high density poly(ethene) Type of poly(ethene) that is rigid and not transparent. The polymer molecules can pack closely together because they do not contain side branches.

hormone Chemical made by a gland and carried in the blood to its target organ(s).

hydrated Describes a salt that contains water of crystallisation in its structure, such as $CuSO_4.5H_2O$.

hydrogen ion Hydrogen atom that has lost its electron, H+.

hydroxide ion Ion with the formula OH-. It makes solutions alkaline.

inheritance factor Term used by Mendel. We now call this a gene.

insulating material Does not allow electrons to move easily through it.

intermolecular forces Relatively weak forces between a molecule and its neighbours.

ion Atom or group of atoms that has gained or lost electrons, so is electrically charged.

ionic bonding Strong electrostatic forces between oppositely charged ions. Ionic bonding acts in all directions.

ionic compound Compound made up of positively and negatively charged ions.

ionic equation Equation that summarises a reaction between ions by showing only the ions that take part in the reaction.

ionising radiation Alpha, beta, and gamma radiation that removes electrons from the atoms of the material it passes through.

isolation Separation of two populations of a species so that they cannot interbreed, for example by a geographical barrier such as an ocean or mountain range.

isomerase Type of enzyme that rearranges the atoms in a molecule.

isotope Atom with the same number of protons but a different number of neutrons.

joule Unit of energy or work done.

kinetic energy Energy any object has when it is moving.

lactic acid Chemical made from the incomplete breakdown of glucose during anaerobic respiration.

law of conservation of momentum Law that states that momentum is conserved in collisions. Total momentum of a system before a collision or explosion is the same as the total momentum of the system after a collision or explosion.

leaf Plant organ specialised for photosynthesis.

light-dependent resistor Special type of resistor whose resistance decreases when the intensity of light falling on it increases.

light-emitting diode Diode that emits light when an electric current flows through it.

limit of proportionality Point beyond which the extension of an object is no longer directly proportional to the force applied.

limiting factor Factor such as carbon dioxide level, light, or temperature, that will affect the rate of photosynthesis if it is in short supply. Increasing the limiting factor will increase the rate of photosynthesis.

lipase Enzyme that catalyses the breakdown of fats (lipids) to fatty acids and glycerol.

live Pin/wire that transfers energy to an appliance.

low density poly(ethene) Type of poly(ethene) that is flexible and transparent. The polymer molecules contain side branches that prevent them packing closely together.

lubricant Substance that helps moving parts slide over each other easily.

macromolecule Three-dimensional network of atoms that are joined together by covalent bonds in a regular pattern.

mass Amount of matter in an object. Measured in kilograms.

mass number The total number of protons and neutrons in the nucleus of an atom.

mass spectrometer Instrument that identifies substances quickly and accurately, and that can give their relative molecular masses.

maximum theoretical yield Amount of product, calculated from the reaction equation, that would be obtained if none was lost when the reaction was carried out.

mean Average of a collection of data.

median Middle value of a collection of data.

meiosis Type of cell division that occurs to form sex cells, resulting in four genetically different cells.

metal halide A compound made up of a metal and a halogen.

metallic bonding Forces of attraction between positively charged metal ions and delocalised electrons in a metal.

microscope Instrument used to view small objects. A light microscope magnifies up to about 1500 times; an electron microscope gives higher magnifications.

mitochondria Structures in animal and plant cells where aerobic respiration takes place.

mitosis Type of cell division that occurs in body cells, resulting in two genetically identical cells.

mobile phase In chromatography, the solvent that carries chemicals from a sample through a chromatography column or along a piece of chromatography paper.

mode Most popular value of a collection of data.

mole The relative formula mass of a substance in grams.

molecular formula Description of a compound or an element that uses symbols for atoms. It shows the relative number of atoms of each type in the substance.

molecular ion peak Peak on a mass spectrograph whose mass represents the relative formula mass of the molecule being analysed.

molecule Particle made up of two or more atoms chemically bonded together.

momentum Mass of an object × velocity.

monohybrid inheritance Inheritance pattern of a single characteristic, determined by one gene (that may have two alleles).

mucus Slimy substance made in special cells/glands.

multicellular Being built of many cells, all working together as an organism.

mutation Change in the structure of a gene. A mutation may result in the gene coding for a different protein/ characteristic.

nanoparticle Tiny particle made up of only a few hundred atoms. It measures between 1 nanometre and 100 nanometres across.

nanoscience Study of nanoparticles.

nanotube Sheets of carbon atoms arranged in hexagons that are wrapped around each other to form a cylinder with a hollow core.

nebula Huge cloud of gas and dust in space (mainly hydrogen).

neutral Pin/wire that completes the circuit.

neutralisation reaction Reaction in which hydrogen ions react with hydroxide ions to produce water.

neutron Sub-atomic particle found in the nucleus of an atom. It has no charge and a relative mass of 1.

neutron star A small, dense object that remains after a supernova.

newton Unit of force.

nuclear fission Splitting a nucleus into two smaller nuclei, releasing energy.

nuclear fusion Fusing two atoms to form a single, larger nucleus, releasing energy.

nucleus Structure inside a cell that controls the cell's activities. It contains chromosomes made of DNA.

nucleus Relatively heavy central part of an atom, made up of protons and neutrons.

optimum Best.

organ Collection of different tissues working together to perform a function within an organism; examples include the stomach in an animal and the leaf in a plant.

organ system Collection of different organs working together to perform a major function within an organism; an example is the digestive system in an animal.

oscilloscope Device that can be used to determine the frequency and potential difference of an a.c. supply.

oxidation Process in which oxygen is added to an element or compound, or by which an atom or ion loses electrons.

oxide ion An atom of oxygen that has gained two electrons to form an ion with two negative charges, O^{2-}.

oxygen debt Lack of oxygen in muscle cells. Oxygen is needed to oxidise lactic acid in the muscle to carbon dioxide and water.

pair of forces Two equal and opposite forces produced when objects interact.

palisade layer A layer of tall columnar cells containing chloroplasts, where the majority of photosynthesis occurs in a leaf.

palisade mesophyll cells Tall columnar cells containing chloroplasts, where the majority of photosynthesis occurs in a leaf.

paper chromatography Method of separating substances in a mixture by allowing a solution of the mixture to flow along a sheet of special paper. If the substances in the mixture travel at different speeds, they will be separated.

parallel circuit Circuit with components connected side by side so that there is more than one path to take around the circuit.

percentage by mass The percentage by mass of an element in a compound is equal to the number of grams of the element in 100 grams of the compound.

percentage yield Actual yield × 100 ÷ maximum theoretical yield.

permanent vacuole Fluid-filled cavity in plant cells.

pH scale Scale that describes the concentration of hydrogen ions in a solution.

phloem Plant tissue made up of living cells that has the function of transporting food substances through the plant.

photosynthesis Process by which plants build carbohydrates from carbon dioxide and water, using sunlight energy.

plutonium 239 Isotope of plutonium that can be used to generate heat in nuclear power stations.

population The number of organisms of a species in a given area.

potential difference Difference in voltage between two points. Measured in volts. Also a measure of the difference in energy carried by electrons between two points.

power Work done (or energy transferred) in a given time. Measured in watts.

precipitate Suspension of small solid particles, spread throughout a liquid or solution.

precipitation reaction Reaction in which a precipitate forms.

protease Enzyme that catalyses the breakdown of proteins to amino acids.

proteins Large molecules (polymers) made of many amino acids joined together. Proteins have many functions, including structural (as in muscle), hormones, antibodies, and enzymes.

proton Sub-atomic particle found in the nucleus of an atom. It has a charge of +1 and a relative mass of 1.

protostar Ball of hot, dense gas, on its way to becoming a star.

radioactive decay Breakdown of an unstable nucleus, giving out alpha particles, beta particles, or gamma rays.

random Having no set pattern in which atoms will decay.

rate of photosynthesis How quickly a plant is photosynthesising. The rate is affected by factors including carbon dioxide levels, light, and temperature.

rate of reaction Amount of product made ÷ time, or the amount of reactant used ÷ time.

reaction time Time taken from seeing a hazard to starting to press the brake pedal.

recessive Visible characteristic that is only present in an organism if two alleles of the gene are present.

red giant The result when a smaller star begins to die. It expands and cools, forming a red giant.

red super giant When a large star begins to die it expands and cools, forming a gigantic red super giant.

reduction Process in which oxygen is removed from a compound, or an atom or ion gains electrons.

regenerative braking A type of vehicle braking that converts some of the kinetic energy into a useful form (eg chemical energy in batteries).

relationship Interaction between different species living together in the same area, such that one species affects another. An example is a predator–prey relationship.

relative atomic mass The relative atomic mass of an element compares the mass of atoms of the element with the mass of atoms of the ^{12}C isotope. It is an average of the values for the isotopes of the element, taking into account their relative amounts. Its symbol is A_r.

relative formula mass The relative formula mass of a substance is the mass of a formula unit of that substance compared to the mass of a ^{12}C carbon atom. It is worked out by adding together all the A_r values for the atoms in the formula. Its symbol is M_r.

repel Push away.

residual current circuit breaker Circuit breaker that detects a difference in current between the live and neutral wire.

resistance Measure of how difficult it is for electrons to pass through a component. Measured in ohms (Ω).

resistive force Force in the opposite direction to the direction of motion of an object.

resistor Circuit component that reduces the current flowing in a circuit.

respiration Process by which living things release energy from carbohydrates, also producing water and carbon dioxide.

resultant force Single force that would have the same overall effect of all the forces on an object combined.

retention time In chromatography, the time it takes for a chemical in a mixture to move through the stationary phase.

reversible reaction Reaction in which the products of a reaction can react to produce the original reactants.

salt Compound that contains metal ions and that can be made from an acid.

sampling Counting a small number of a large total population in order to study its distribution.

series Circuit with components that are connected end to end so there is only one way around the circuit.

sex chromosomes Chromosomes involved with determining the sex of an individual. In humans if you have two X chromosomes you are female; if you have one X and one Y chromosome you are male.

shape memory alloy Alloy that, when bent or twisted, keeps its new shape, but when heated will return to its original shape.

soluble Substance that dissolves in a given solvent.

specialised Specialised cells have a structure well suited to their function.

speciation Separate evolution of two populations of the same species, to form two separate species.

specific Enzymes are specific; they act on only one substrate.

speed Distance moved ÷ time taken. How fast something is moving.

speed camera Camera used to find the speed of a vehicle.

starch Large insoluble carbohydrate molecule, stored in plant cells.

static electricity Build up of charge from electrons that have been moved to or from an insulator.

stationary Not moving, at rest.

stationary phase In chromatography, the medium through which the mobile phase passes.

stem cell Undifferentiated cell that can divide by mitosis and is capable of differentiating into any of the cell types found in that organism.

sterilising Using ionising radiation (usually gamma rays) to kill microorganisms on medical equipment.

stomata Pores on the surface of a leaf that allow water, carbon dioxide, and oxygen to move in and out of the leaf.

stopping distance Sum of the thinking distance and braking distance.

sub-atomic particle Describes particles from which atoms are made, including protons, neutrons, and electrons.

substrate Substance acted upon by an enzyme in a chemical reaction. The substrate molecules are changed into product molecules.

successful collision Collision between reactant particles that results in a reaction.

supernova What happens when a massive star explodes at the end of its life.

surface area In a chemical reaction involving a solid, the area of a solid that is in contact with the other reactants.

switch Circuit component that can switch the circuit on or off by closing or making a gap in the circuit.

terminal velocity Maximum velocity of an object when the forces on it are balanced. Usually applied to objects falling under gravity.

thermistor Special type of resistor whose resistance decreases as the temperature increases.

thermosetting polymer Polymer that does not melt.

thermosoftening polymer Polymer that softens easily on heating and that can be moulded into new shapes.

thinking distance Distance travelled by a vehicle in the reaction time.

time base Time interval represented by each horizontal square on an oscilloscope.

time period Time for one complete cycle of potential difference for an alternating current.

tissue Group of cells of similar structure and function working together, such as muscle tissue in an animal and xylem in a plant.

tract A system of tubes in the body, such as the digestive tract or respiratory tract.

uranium 235 Isotope of uranium that can be used to generate heat in nuclear power stations.

variable resistor Resistor whose resistance can be changed.

velocity How fast something is moving in a certain direction.

velocity–time graph Graph with velocity on the y-axis and time on the x-axis. It shows the velocity of an object at a particular moment.

voltage *See* potential difference.

watt Unit of power.

weight Force on an object due to the force of gravity on a planet.

white dwarf White-hot core left after the outer layers of a red giant break away.

work done Energy transferred, measured in joules. Work done is equal to force applied × distance moved in the direction of the force.

xylem Plant tissue made up of dead cells that has the function of transporting water and dissolved substances through the plant.

yield Amount of the required product made in a reaction.

zygote (Usually) diploid cell resulting from the fusion of an egg and a sperm.

Index

Reference material

Periodic table

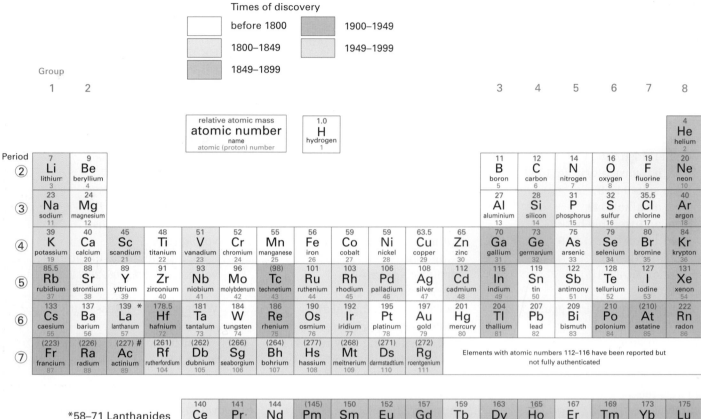

Times of discovery

before 1800		1900–1949
1800–1849		1949–1999
1849–1899		

Reactivity series of metals

Potassium	most reactive
Sodium	
Calcium	
Magnesium	
Aluminium	
Carbon	
Zinc	
Iron	
Tin	
Lead	
Hydrogen	
Copper	
Silver	
Gold	
Platinum	least reactive

(elements in italics, though non-metals, have been included for comparison)

Formula of some common ions

Name	Formula	Name	Formula
Hydrogen	H^+	Chloride	Cl^-
Sodium	Na^+	Bromide	Br^-
Silver	Ag^+	Fluoride	F^-
Potassium	K^+	Iodide	I^-
Lithium	Li^+	Hydroxide	OH^-
Ammonium	NH_4^+	Nitrate	NO_3^-
Barium	Ba^{2+}	Oxide	O^{2-}
Calcium	Ca^{2+}	Sulfide	S^{2-}
Copper(II)	Cu^{2+}	Sulfate	SO_4^{2-}
Magnesium	Mg^{2+}	Carbonate	CO_3^{2-}
Zinc	Zn^{2+}		
Lead	Pb^{2+}		
Iron(II)	Fe^{2+}		
Iron(III)	Fe^{3+}		
Aluminium	Al^{3+}		

Equations

$a = \dfrac{F}{m}$ or $F = M \times a$	F is the resultant force in newtons, N m is the mass in kilograms, kg a is the acceleration in metres per second squared, m/s²
$a = \dfrac{v - u}{t}$	a is the acceleration in metres per second squared, m/s² v is the final velocity in metres per second, m/s u is the initial velocity in metres per second, m/s t is the time taken in seconds, s
$W = m \times g$	W is the weight in newtons, N m is the mass in kilograms, kg g is the gravitational field strength in newtons per kilogram, N/kg
$F = k \times e$	F is the force in newtons, N k is the spring constant in newtons per metre, N/m e is the extension in metres, m
$W = F \times d$	W is the work done in joules, J F is the force applied in newtons, N d is the distance moved in the direction of the force in metres, m
$P = \dfrac{E}{t}$	P is the power in watts, W E is the energy transferred in joules, J t is the time taken in seconds, s
$E_p = m \times g \times h$	E_p is the change in gravitational potential energy in joules, J m is the mass in kilograms, kg g is the gravitational field strength in newtons per kilogram, N/kg h is the change in height in metres, m
$E_k = \dfrac{1}{2} \times m \times v^2$	E_k is the kinetic energy in joules, J m is the mass in kilograms, kg v is the speed in metres per second, m/s
$p = m \times v$	p is the momentum in kilograms metres per second, kg m/s m is the mass in kilograms, kg v is the velocity in metres per second, m/s
$I = \dfrac{Q}{t}$	I is the current in amperes (amps), A Q is the charge in coulombs, C t is the time in seconds, s
$V = \dfrac{W}{Q}$	V is the potential difference in volts, V W is the work done in joules, J Q is the charge in coulombs, C
$V = I \times R$	V is the potential difference in volts, V I is the current in amperes (amps), A R is the resistance in ohms, Ω
$p = \dfrac{E}{t}$	P is power in watts, W E is the energy in joules, J t is the time in seconds, s
$P = I \times V$	P is power in watts, W I is the current in amperes (amps), A V is the potential difference in volts, V
$E = V \times Q$	E is the energy in joules, J V is the potential difference in volts, V Q is the charge in coulombs, C

Fundamental physical quantities	
Physical quantity	**Unit(s)**
length	metre (m) kilometre (km) centimetre (cm) millimetre (mm)
mass	kilogram (kg) gram (g) milligram (mg)
time	second (s) millisecond (ms)
temperature	degree Celsius (°C) kelvin (K)
current	ampere (A) milliampere (mA)
voltage	volt (V) millivolt (mV)

Derived quantities and units	
Physical quantity	**Unit(s)**
area	cm^2; m^2
volume	cm^3; dm^3; m^3; litre (l); millilitre (ml)
density	kg/m^3; g/cm^3
force	newton (N)
speed	m/s; km/h
energy	joule (J); kilojoule (kJ); megajoule (MJ)
power	watt (W); kilowatt (kW); megawatt (MW)
frequency	hertz (Hz); kilohertz (kHz)
gravitational field strength	N/kg
radioactivity	becquerel (Bq)
acceleration	m/s^2; km/h^2
specific heat capacity	J/kg°C
specific latent heat	J/kg

Electrical symbols

junction of conductors		ammeter	(A)	diode		capacitor	
switch		voltmeter	(V)	electrolytic capacitor		relay	NO COM NC
primary or secondary cell		indicator or light source		LDR		LED	
battery of cells	or			thermistor		NOT gate	NOT
power supply		motor	(M)	AND gate	AND	OR gate	OR
fuse		generator	G	NOR gate	NOR	NAND gate	NAND
fixed resistor		variable resistor					

Acknowledgements

The publisher and authors would like to thank the following for their permission to reproduce photographs and other copyright material:

p8T Andrew Lambert Photography/SPL; p8B 81a/Alamy; p13 Carmen Martinez Banus/Istockphoto; p14M Photo Researchers/SPL; p14R Omikron/SPL; p14L Dr Gopal Murti/SPL; p15 John Durham/SPL; p16 Dr. Terry Beveridge, Visuals Unlimited/SPL; p17 Science VU, Visuals Unlimited/SPL; p18T Power And Syred/SPL; p18MT Pasieka/SPL; p18M Michael Abbey/SPL; p18MB Eye Of Science/SPL; p18B Steve Gschmeissner/SPL; p19T Dr David Furness, Keele University/SPL; p19MT Steve Gschmeissner/SPL; p19MB Dr David Furness, Keele University/SPL; p19B J.C. Revy, Ism/SPL; p20L Andrew Lambert Photography/SPL; p20R Andrew Lambert Photography/SPL; p22T Science VU, Visuals Unlimited/SPL; p22M Dr. Richard Kessel & Dr. Gene Shih, Visuals Unlimited/SPL; p22B CNRI/SPL; p23 Dr. Richard Kessel & Dr. Gene Shih, Visuals Unlimited/SPL; p25 Steve Gschmeissner/SPL; p26 Gavin Kingcome/SPL; p27 Power And Syred/SPL; p30R Veronique Leplat/SPL; p30L Power And Syred/SPL; p31 Carlos Munoz-Yague/Eurelios/SPL; p32 Scott Sinklier/Agstockusa/SPL; p34 Jennifer Fry/SPL; p36BL Martyn F. Chillmaid/SPL; p36BR Martyn F. Chillmaid/SPL; p36T Paul Harcourt Davies/SPL; p38 Colin Cuthbert/Newcastle University/SPL; p39 George Steinmetz/SPL; p45 PH. Plailly/Eurelios/SPL; p46L John Bavosi/SPL; p46R Tim Vernon/SPL; p47 Laguna Design/SPL; p50 Roger Harris/SPL; p52R Cordelia Molloy/SPL; p52BL Power And Syred/SPL; p52TL Maximilian Stock Ltd/SPL; p53 Manfred Kage/SPL; p54 AJ Photo/SPL; p55T Ashley Cooper, Visuals Unlimited/SPL; p55B Jonathan Hordle/Rex Features; p56T Tony Craddock/SPL; p56B Joe Mcdonald, Visuals Unlimited/SPL; p57 Manfred Danegger/Okapia/SPL; p58T Dr. Fred Hossler, Visuals Unlimited/SPL; p58B GustoImages/SPL; p59 CNRI/SPL; p60 Friedrich Saurer/SPL; p61 Stigur Karlsson/Istockphoto; p62 Pr. G Gimenez-Martin/SPL; p63 Herve Conge, ISM/SPL; p65B Eye Of Science/SPL; p65T Adrian T. Sumner/SPL; p66B Bruno Petriglia/SPL; p66T James King-Holmes/SPL; p70L SPL; p70R Gary Parker/SPL; p72 Jacopin/SPL; p74L Larry Dunstan/SPL; p74R CNRI/SPL; p76 Professor Miodrag Stojkovic/SPL; p77 Martin Shields/SPL; p78T Frederick R. McConnaughey/SPL; p78MT Sinclair Stammers/SPL; p78MB Pasieka/SPL; p78B Olivier Darmon/Jacana/SPL; p80 Christian Darkin/SPL; p81L Photo Researchers/SPL; p81TR Dan Sams/SPL; p81BR Simon Fraser/SPL; p82 Science Source/SPL; p83 David Gifford/SPL; p89 Dr Peter Harris/SPL; p90TL Tim Scrivener/Rex Features; p90TR Dmitri Melnik/Shutterstock; p90BL Martyn F. Chillmaid/SPL; p90BR Georgette Douwma/SPL; p92 Springfield Gallery/Fotolia; p94 Shane White/Shutterstock; p95 John Chumack/SPL; p96T Hulton Archive/Stringer/Getty images; p96B Charles D. Winters/SPL; p98 Martin Lovatt/Istockphoto; p99 Tom Watkins/Rex Features; p100 Charles D. Winters/SPL; p101 Eliza Snow/Istockphoto; p102R Eye of Science/SPL; p102L Oleg Mitiukhin/Istockphoto; p104R Radius Images/Photolibrary; p104L David Hoffman Photo Library/Alamy; p105 Ryan Balderas/Istockphoto; p106 Kevin Britland/Shutterstock; p107 US Air Force/SPL; p108R Fertnig/Istockphoto; p108L Brent Danley; p109 Tony Mcconnell/SPL; p110L Tyler Olson/Shutterstock; p110R Steve Meddle/Rex Features; p112 Hank Morgan/SPL; p114 Philippe Plailly/SPL; p116 Peter Menzel/SPL; p118 Tracy Hebden/Alamy; p119 Simon Fraser/SPL; p120 Ivanru/Dreamstime; p121 Martyn F. Chillmaid/SPL; p122L Martyn F. Chillmaid/SPL; p122M Charles D. Winters/SPL; p122R Charles D. Winters/SPL; p123 Andrew Lambert Photography/SPL; p129 Javier Trueba/MSF/SPL; p130L Daniel Gale/Shutterstock; p130R Photos.com; p132R Cordelia Molloy/SPL; p132L Martyn F. Chillmaid/SPL; p135 Charles D. Winters/SPL; p136R Photos.com; p136L Martyn F. Chillmaid/SPL; p138 Charles D. Winters/SPL; p139 Charles D. Winters/SPL; p140BL Khanh Trang/Istockphoto; p140BM Jenny Swanson/Istockphoto; p140BR Peter Elvidge/Istockphoto; p140T Ruslan Gilmanshin/Istockphoto; p141L Nick Free/Istockphoto; p141R Michael Durham/Getty; p142TR Cordelia Molloy/SPL; p142BR Charles D. Winters/SPL; p142L Andrew Lambert Photography/SPL; p143T Carolyn A. McKeone/SPL; p143B Pablo del Rio Sotelo/Istockphoto; p144R Dr Jeremy Burgess/SPL; p144L SPL; p146 Marcus Leith, London. Image courtesy of Corvi-Mori, London; p148 Sidney Moulds/SPL; p150 Lawrence Migdale/SPL; p151 Charles D. Winters/SPL; p152TL Alex Segre/Rex Features; p152B David Partington/Istockphoto; p152TR Mint Photography/Alamy; p154T Solent News & Photo Agency/Rex Features; p154B JW.Alker/Photolibrary; p156R Adam Hart-Davis/SPL; p156TL Joe Gough/Istockphoto; p156BL Cristian Lucaci/Istockphoto; p159 James Holmes, Hays Chemicals/SPL; p165 Nürburgring Automotive GmbH/Fotoagentur Urner; p166 Jim Grossman/NASA; p167 Heathcliff O'Malley/Rex Features; p168 Akihiro Sugimoto/Photolibrary; p169 Ian Cuming/SPL; p170R Andrew Wong/Getty Images Sport/Getty Images; p170L Cordelia Molloy/SPL; p171 Edward Shaw/Istockphoto; p174B Kim Kirby/Photolibrary; p174T Alan & Sandy Carey/SPL; p175 nicmac.ca/Fotolia; p179 Sankei/Getty Images News/Getty Images; p180 Barry Phillips/Evening Standard/Rex Features; p181L Ken McKay/Rex Features; p181R James D. Morgan/Rex Features; p185 Patrick Eden/Alamy; p186 Martyn F. Chillmaid/SPL; p188 Orange Line Media/Shutterstock; p189B NASA/SPL; p189T Paul Bernhardt/Alamy; p190R Stephen Hird/Reuters; p190L Andrew Lambert Photography/SPL; p191 Kondrashov Mikhail Evgenevich/Shutterstock; p193 Yuriko Nakao/Reuters; p194 Joshua Hodge Photography/Istockphoto; p201 Keith Kent/SPL; p203R Leslie Banks/Istockphoto; p203L Charles D. Winters/SPL; p206T1 Andrew Lambert Photography/SPL; p206T2 Paul Reid/Shutterstock; p206T3 Chris Hutchison/Istockphoto; p206T4 Andrew Lambert Photography/SPL; p206T5 Andrew Lambert Photography/SPL; p206T6 Trevor Clifford Photography/SPL; p206T7 Doug Martin/SPL; p206T8 Webking/Istockphoto; p206T9 Алексей Брагин/Istockphoto; p206T10 Andrew Lambert Photography/SPL; p206T11 Martyn F. Chillmaid/SPL; p206T12 Martyn F. Chillmaid/SPL; p207 Trevor Clifford Photography/SPL; p208 Trevor Clifford Photography/SPL; p210 Doug Martin/SPL; p212 Doug Martin/SPL; p213 Trevor Clifford Photography/SPL; p214 Pali Rao/Istockphoto; p215 Martyn F. Chillmaid/SPL; p216 Martyn F. Chillmaid/SPL; p217T Martyn F. Chillmaid/SPL; p217B Eyewave/Istockphoto; p218 Studiomode/Alamy; p219 Andrew Lambert Photography/SPL; p220TL Krasowit/Shutterstock; p220TR Jorge Farres Sanchez/Dreamstime; p220B oksana2010/Shutterstock; p221 David J. Green/Alamy; p222 Andrew Lambert Photography/SPL; p227L Philippe Psaila/SPL; p227R SPL; p229 Patrick Landmann/SPL; p231 Prof. J. Leveille/SPL/SPL; p233 EFDA-JET/SPL; p234 NASA; p235 NASA.

Illustrations by Wearset Ltd, HL Studios, James Stayte.

Although we have made every effort to trace and contact all copyright holders before publication this has not been possible in all cases. If notified, the publisher will rectify any errors or omissions at the earliest opportunity.

OXFORD

UNIVERSITY PRESS

Great Clarendon Street, Oxford OX2 6DP

Oxford University Press is a department of the University of Oxford.
It furthers the University's objective of excellence in research,
scholarship, and education by publishing worldwide in

Oxford New York

Auckland Cape Town Dar es Salaam Hong Kong Karachi
Kuala Lumpur Madrid Melbourne Mexico City Nairobi
New Delhi Shanghai Taipei Toronto

With offices in
Argentina Austria Brazil Chile Czech Republic France Greece
Guatemala Hungary Italy Japan Poland Portugal Singapore
South Korea Switzerland Thailand Turkey Ukraine Vietnam

Oxford is a registered trade mark of Oxford University Press
in the UK and in certain other countries.

© Oxford University Press 2011

The moral rights of the authors have been asserted

Database right Oxford University Press (maker)

First published 2011

All rights reserved. No part of this publication may be reproduced,
stored in a retrieval system, or transmitted, in any form or by any means,
without the prior permission in writing of Oxford University Press, or as
expressly permitted by law, or under terms agreed with the appropriate
reprographics rights organization. Enquiries concerning reproduction
outside the scope of the above should be sent to the Rights Department,
Oxford University Press, at the address above.

Cover image courtesy of Gusto Images/Science Photo Library

You must not circulate this book in any other binding or cover
and you must impose this same condition on any acquirer.

British Library Cataloguing in Publication Data

Data available

ISBN 978-0-19-913588-2

10 9 8 7 6 5 4 3 2 1

Printed in Great Britain by Bell and Bain, Glasgow

Paper used in the production of this book is a natural, recyclable product
made from wood grown in sustainable forests. The manufacturing process
conforms to the environmental regulations of the country of origin.

WALWORTH ACADEMY
SHORNCLIFFE ROAD
LONDON
SE1 5UJ

Mixed Sources
Product group from well-managed
forests and other controlled sources
www.fsc.org Cert no. TT-COC-002769
© 1996 Forest Stewardship Council

FSC